Charles Joseph Gahan

A List of the Longicorn Coleoptera

Collected by Signor Fea in Burma and the adjoining regions, with

descriptions of the new genera and species

Charles Joseph Gahan

A List of the Longicorn Coleoptera
Collected by Signor Fea in Burma and the adjoining regions, with descriptions of the new genera and species

ISBN/EAN: 9783337240707

Printed in Europe, USA, Canada, Australia, Japan

Cover: Foto ©Andreas Hilbeck / pixelio.de

More available books at **www.hansebooks.com**

VIAGGIO DI LEONARDO FEA
IN BIRMANIA E REGIONI VICINE

LVI.

A LIST OF THE LONGICORN COLEOPTERA

COLLECTED BY SIGNOR FEA

IN BURMA AND THE ADJOINING REGIONS

WITH DESCRIPTIONS

OF THE NEW GENERA AND SPECIES

by CHARLES J. GAHAN, M. A.

of the British Museum (Nat. History)

GENOVA
TIPOGRAFIA DEL R. ISTITUTO SORDO-MUTI
1894

VIAGGIO DI LEONARDO FEA
IN BIRMANIA E REGIONI VICINE

LVI.

A List of the *Longicorn Coleoptera* collected by Signor FEA in Burma and the adjoining regions, with descriptions of the new Genera and species by CHARLES J. GAHAN, M. A., of the British Museum (Nat. History).

(*Plate I*).

The Longicorn Coleoptera recorded in the following list amount to a total of about 240 species. Of this number 90 are now described for the first time and, with 3 previously described by Dr Gestro, give 93 as the number of new species contained in the collections made by Signor Fea. But of these not a few were already represented in other collections by specimens which had been brought from Burma, or, as was more often the case, from North India. Where it seemed desirable, as tending to throw some light upon the distribution of the species, I have added after each the other localities from which it was known, in addition to those indicated by Signor Fea's specimens. A comparison of these localities will show that the Longicorn fauna of Burma has very much in common with that of Nepal, Sikkim and Assam in Northern Hindostan. This is more particularly true, perhaps, of mountain species, some of which range as far south even as Sumatra and Java. There is less resemblance between the Longicorns of South India and Burma, though a few species are common to both which have not so far been recorded from North India. Among the species that appear to be confined to Burma may be more especially mentioned those of the fine genus *Arctolamia* described by Dr Gestro, and some large species which I have placed in the genus *Niphona*. Fea's collections have furnished two species of *Atimura*, one of which is indistinguishable from an Australian species described by Pascoe. Of the remaining species of *Atimura* two are from the Malayan region, and one from Japan. It is probable that this genus is represented by a much larger number of species, but that owing to their small size and the protection afforded them by their great resemblance to little bits of sticks they have to a great extent escaped the notice of collectors. The new species, for which I have proposed the genus *Estigmenida*, is also remarkable for its striking resemblance to a common Oriental Hispid *Estigmena chinensis*, Hope. The majority of the new species belong to the groups *Clytinae*, *Mesosinae*, *Niphoninae* and *Obereinae*, and it may be found that, owing to their small size and the difficulty of identifying species in these perplexing groups, some of them have already been described. In identifying some of the more difficult members of the group *Callichrominae* I was indebted to the late Mr Bates for his usual kind assistance. While to Mr Pascoe and to Mr René Oberthür I have to render thanks, for the facilities they gave me in examining types from their collections.

PRIONIDAE.

1. **Cyrtognathus** (*Baladeva*) **Walkeri**, Waterh., Trans. Ent. Soc. Lond. vol. II, p. 226, pl. 21, fig. 1.

Carin Mts. district of Chebà; Alt. 900-1100 metres. Cambodia (Brit. Mus. collection).

2. **Cyrtognathus** (*Paraphrus*) **granulosus**, Thoms., Essai d'une Class. des Cérambycides, p. 329.

Bhamò, Teinzò (Upper Burma) and Thagatà a village, on Mt. Mooleyit 400 or 500 m. above the sea. Occurs also in Siam and in N. India.

3. **Cyrtognathus** (*Paraphrus*) sp.

Carin Mts. (Asciuii Chebà). Alt. 1200-1300 m. One female example.

4. **Cyrtognathus** (*Cyrtosternus*) **indicus**, Hope, Gray's Zoological Miscellany (1831), p. 27.

Teinzò in Upper Burma and Carin Mts., 400-500 m. altit.; also North India (Brit. Mus. coll.).

5. **Ancyloprotus bigibbosus**, White, Cat. Longicornia, Brit. Mus. p. 19.

Palon; Carin Mts., district of Chebà; Alt. 900-1100 metres. Occurs also in Assam.

6. **Macrotoma Fisheri**, C. O. Waterh., Ann. & Mag. Nat. Hist. Ser. 5, vol. XIV (1884), p. 382.

Carin, Mts. Chebà; Alt. 900-1100 m.

7. **Remphan Hopei**, Waterh., Trans. Ent. Soc. Lond., vol. 1, p. 67; pl. 8, fig. 1.

Palon in Pegu. The species occurs also in Penang, Borneo and the Andaman Islands.

8. **Sarmydus subcoriaceus**, Hope. — *Prionus subcoriaceus*, Hope, in Gray's Zool. Miscellany (1831), p. 27. — *Sarmydus antennatus*, Pasc., Ann. & Mag. Nat. Hist., Ser. 3, vol. XIX, p. 410; Trans. Ent. Soc. Lond., Ser. 3, vol. III, p. 677, pl. 24, fig. 2 (σ).

Catcin Cauri in Upper Burma. One female example. This species occurs also in Nepal, North India; in the Andaman Islands (*Roepstorff*) and in Borneo (*Wallace*).

9. **Aegosoma marginale**, Fabr. — *Cerambyx marginalis*, Fabr. Syst. Ent., p. 169.

Teinzò in Upper Burma. One example.

10. **Aegosoma sulcipenne**, White, Cat. Longic. Brit. Mus., p. 30.

Teinzò in Upper Burma. Occurs in North India.

11. **Aegosoma lacertosum**, Pasc., Ann. & Mag. Nat. Hist., Ser. 3, XIX, p. 413.

One example from Carin Mts. (Chebà): Alt. 900-1100 metres. The species is represented in Mr Pascoe's collection by a single specimen from Sylhet. This specimen does not differ by any very important characters from the unique type of *Aegosoma costipenne*, White (*Megopis*) (Longic. Brit. Mus., p. 28, pl. 2, fig. 2) which also came from Sylhet in Assam. The question as to whether these differences are specific or merely individual must, however, be left in abeyance. It can only be settled by the examination of a larger series of examples.

12. **Cyrtonops punctipennis**, White, Cat. Long. Brit. Mus., p. 32, pl. 2, fig. 3 (♀).

Bhamó in Upper Burma; also North India (Brit. Mus. Coll.).

CERAMBYCIDAE.

13. **Tetraommatus callidioides**, Pasc. var., Trans. Ent. Soc. Lond., Ser. 2, vol. IV (1857), p. 98, pl. 23, fig. 6; id. Ser. 3, vol. III, p. 502.

Mt. Heanlain in Upper Burma. One example.

In the single example taken, the front and the upper side of the head are dark brown; and the small prothoracic spines are somewhat more distinct than in the other examples of *T. callidioides* that I have seen. It ought perhaps to be referred to *T. nigriceps*, Pasc., but the characters (with the single exception of the colour of the head) by which M.ʳ Pascoe has distinguished the latter species do not well apply to the present variety.

14. **Tetraommatus insignis**, sp. n. (*Pl.* I, *fig.* 1). *Ferrugineo-testaceus, sparsim pubescens; prothorace antice posticeque transversim sulcato, sulcis infuscatis, lateraliter utrinque tuberculo parvo acuto armato, basi constricto; dorso prope basin longitudinaliter haud profunde bisulcato: elytris dense punctatis, flavo-testaceis, pone medium plaga fusca signatis; corpore subtus pedibusque nonnihil piceo-rufescentibus. Long.* 10-11 *mm.*

Thagatá in Tenasserim; one example. The British Museum collection contains examples from the Andaman Islands.

Head and prothorax dull ferruginous-red. The length of the prothorax to its median breadth in the ratio of 3 to 2. The prothorax is closely but somewhat obscurely punctured; its sides are slightly rounded, and each is armed with a small but sharp and distinct tubercle; the constricted basal part of the prothorax is preceded by a slightly arcuate transverse groove; there is a feebler groove close to the anterior margin, and on the posterior part of the disk there are two shallow longitudinal grooves which pass in front into a broad shallow longitudinal depression. Both the transverse and longitudinal grooves are more or less infuscate. Scutellum testaceous. Elytra thickly punctured; yellowish-testaceous, with a brownish patch, formed by two oblique bands which pass outwards and backwards from the suture on each side between the middle and posterior third, and by the dark borders of that portion of the suture which lies between these bands: the posterior bands are short, the anterior bands extend farther outwards, and, near the margin, take an almost directly transverse direction.

15. **Xystrocera globosa**, Oliv., Ent. IV, no. 67, p. 27, pl. 12, fig. 81.

Catcin Cauri in Burma; and at Carin, district of Ghecù, alt. 1300-1400 metres.

16. **Xystrocera festiva**, Thoms., Essai d'une Class. des Cérambycides, p. 251.

Carin, district of Chebà: alt. 900-1100 metres.

17. **Neocerambyx Paris**, Wied. — *Cerambyx Paris*, Wied., Germ. Mag. Ent. IV (1821), p. 167. — *Cerambyx Brama*, Newm., Ent. Mag., V, p. 493.

Bhamò in Upper Burma: one example. Also North India.

18. **Plocaederus obesus**, Gahan, Ann. & Mag. Nat. Hist., Ser. 6, vol. V, p. 31; id., vol. VI, p. 259. — *Hamaticherus obesus*, Dup., Dej. Cat. 3 edit., p. 347. — *Cerambyx obesus*, Cat. Gemm. & Har., p. 2802.

Myadoung in Upper Burma; Carin Mts. (Asciuii Chebà district) 1200-1300 m.; Mectan in Tenasserim. Occurs also in North India, where it appears to be widely distributed, and in the Andaman Islands.

This species is figured under the name of *Plocaederus pedestris* in "Indian Museum Notes", vol. I, no. 2, pl. V, fig. 4.

19. **Æolesthes sinensis**, Gahan, Ann. & Mag. Nat. Hist., Ser. 6, vol. VI, pp. 252 and 255.

Catcin Cauri in Upper Burma, and Carin Mts., district of Chebà. The Chinese examples upon which this species was founded were in all probability (so Mr Bowring informs me) obtained at Hong Kong or its immediate neighbourhood.

20. **Hoplocerambyx spinicornis**, Newm. — *Hammaticherus spinicornis*, Newm., The Entomologist 1 (1842) p. 245.

Thagatà in Tenasserim; one male example. This species occurs also in the Philippine Islands (type); and in North India — Allahabad and Nepal (Brit. Mus. collection). Examples from Borneo, constituting the *H. morosus* of Pascoe, do not seem to me to offer any definite characters by which they can be distinguished from the above. The species appears to be subject to great variation in size.

21. **Pachydissus** (*Diorthus*) **simplex**, White. — *Hammaticherus simplex*, White, Cat. Longic. Brit. Mus., p. 130 (W. Africa). — *Cerambyx holosericeus*, Oliv. (nec Fabr.) Ent. IV, no. 67, p. 14, pl. 17, fig. 127 (S. India). — *Cerambyx vernicosus*, Pasc., Trans. Ent. Soc. Lond., Ser. 2, vol. V, p. 19 (Ceylon). — *Pachydissus inclemens*, Thoms., Syst. Ceramb., p. 576 (S. India). — *Pachydissus* (*Diorthus*) *simplex*, Gahan, Ann. & Mag. Nat. Hist., Ser. 6, vol. VII, pp. 27 and 31.

Malewoon in Tenasserim. One small example. The localities indicated above, to which may be added Siam and Java, evidence the wideness of distribution of this species. Whether it is an Oriental species, which has been transported to West Africa, or an African species which has been carried to the East, I am unable to decide.

22. **Pachydissus** (*Margites*) **exiguus**, sp. n. *Fuscus, griseo leviter pubescens; antennis pedibusque piceo-rufescentibus, pube grisea tenuiter obtectis; prothorace supra intricate haud fortiter rugoso, lateribus irregulariter rotundatis; antennis quam corpore paullo longioribus, articulis $3°$ et $4°$ apice vix incrassatis, $3°$ quam $1°$ vel $4°$ paullo longiori. Long. 11 mm.*

One example taken at Mandalay. This species resembles *P.* (*Margites*) *egenus*, Pasc. in coloration, but is of much smaller size. The prothorax has a thin unicolorous greyish pubescence.

The third and fourth joints of the antennae are scarcely thickened towards the apex; the fourth is about equal in length to the first; the third and fifth are subequal, each a little longer than the fourth.

23. Dymasius fulvescens, sp. n. (Pl. I, fig. 2).

Piceo-fuscus; capite, prothoracis maculis elytrisque pube fulvo-grisea opaca sat dense obtectis, prothorace quam latitudine longiori, supra intricato-rugoso, lateraliter parum rotundato; elytris postice gradatim angustatis, apicibus truncatis, ad suturam spinosis; corpore subtus, pedibus antennisque griseo leviter pubescentibus, antennis quam corpore paullo longioribus, articulo $3°$ quam $4°$ fere duplo longiori, articulis a $5.°$ ad $10.^{um}$ apice intus spinoso-productis. Long. 20 mm.

Hab. Carin (Ghecù): Alt. 1300-1400 metres. One example.

Pitchy brown. Head with a rather close fulvous-grey pubescence, which is wanting only in the somewhat circular depression limiting the frontal *plaque*, and in the short shallow groove between the antennary tubers. Prothorax longer than broad, slightly rounded at the sides, narrower in front than at the base, intricately and rather strongly rugose above and at the sides; with a fulvous-grey pubescence limited to the anterior border and to a few patches on the anterior part of the disk. Scutellum and elytra clothed with a dense dull fulvous-grey pubescence, which almost completely hides the underlying derm. Elytra gradually narrowed from base to apex, each truncate behind, and spinose at the suture. First joint of the posterior tarsus as long as the two succeeding joints combined.

Antennae ($♀$?) surpassing the elytra by about the last two joints; the fourth joint perceptibly shorter than the scape, and scarcely more than half the length of the third joint; with the joints from the $5.^{th}$ to the $10.^{th}$ spinosely produced at their inner apex, each about equal in length to the $3.^{rd}$; $11.^{th}$ scarcely longer than the $10.^{th}$

24. Mallambyx? sp.

Carin Mts., Asciuii Chebà. One imperfect example.

As the antennae are wanting it is impossible to refer this species with certainty to any particular genus. It seems to me to belong to *Mallambyx* or perhaps to a new genus.

25. **Trachylophus sinensis**, Gahan, Ann. & Mag. Nat. Hist., Ser. 6, vol. II, p. 60.

Carin Mts., district of Ghecù; Alt. 1300-1400 m.; also South China, ? Hong Kong (Brit. Mus. collection). The occurrence of this species as far South as the Carin Mts. makes me inclined to believe that the differences upon which I relied for the separation of examples from Java, under the name of *T. approximator*, may be of an individual rather than of a specific character.

26. **Xoanodera regularis**, Gahan, Ann. & Mag. Nat. Hist., Ser. 6, vol. V, p. 52.

Thagatà in Tenasserim; also occurs in N. India.

27. **Pachylocerus pilosus**, Guér. Men., Icon. Règne Anim. III, p. 230.

Carin, Chebà: Alt. 900-1100 metres.

28. **Stromatium barbatum**, Fabr. — *Callidium barbatum*, Fabr., Syst. Ent., p. 189.

Taken at Rangoon, Tharrawaddy and Prome (Pegu) by Signor Fea; and at Tharrawaddy by Mr Corbett. This widely distributed species seems to be very common throughout nearly the whole of India.

29. **Stromatium asperulum**, White, Cat. Longic. Brit. Mus., p. 300.

Teinzò in Upper Burma. One example. Also occurs in North India, South China, Siam and Penang.

30. **Hesperophanes erosus**, sp. n. *Piceo-brunneus, pube breve griseo-brunnea, supra maculatim disposita, vestitus; capite sat dense pubescente, supra linea media impresso; prothorace supra convexo, obsolete punctulato, maculis longitudinalibus irregularibus griseo-brunneis, dorso antice in medio punctis paucis magnis haud profunde impresso; scutello griseo; elytris crebre fortiterque punctatis, pube griseo-brunnea maculatim disposita, utrisque lineis duabus obsolete elevatis, apicibus rotundatis; corpore subtus, pedibus antennisque leviter pubescentibus; his medium elytrorum paullo excedentibus.* ♀ Long. 23 mm.

One example, taken at Meetan in Tenasserim. Of a reddish brown colour, somewhat darker above. Head with a rather dense greyish brown pubescence; with a median impressed line above. Prothorax convex above and below, strongly rounded at the sides; with some irregular longitudinal spots or patches of greyish brown pubescence, the spaces between which are obsoletely punctured and opaque; with a few large but shallow punctures on the middle of the disk anteriorly. Elytra strongly and very thickly punctured, each with two very feebly raised longitudinal lines; with a short closely laid pubescence disposed in irregular and partly confluent spots or patches; with the naked intervening spaces somewhat glossy; apices obtusely rounded. Body underneath, legs and antennae with a rather thin greyish brown pubescence. Antennae in the female reaching a little beyond the middle of the elytra.

This species is founded upon a single female specimen taken at Meetan in Tenasserim. In the absence of the male sex, it is uncertain whether the species should be referred to *Stromatium* or *Hesperophanes*.

31. **Gnatholea eburifera**, Thom., Class. des Cérambycides, p. 375.

Thagatà in Tenasserim; also Cambodia (type).

32. **Gnatholea simplex**, Gahan, Ann. & Mag. Nat. Hist., Ser. 6, vol. V, p. 53.

Taken by Signor Fea at Prome in Pegu and at Mandalay in Upper Burma; the type specimens in the Brit. Mus. collection are from Darjeeling in N. India. The species has also been captured at Tharawaddy by Mr Corbett.

33. **Nyphasia orientalis**, White. — *Sphaerion orientale*, White, Longic. Brit. Mus., p. 110.

Carin Mts., district of Ghecù; alt. 1300-1400 metres.

34. **Ceresium leucosticticum**, White, Longic. Brit. Mus., p. 245, pl. VI, fig. 1.

Teinzò and Bhamò in Upper Burma; occurs also in Assam.

35. **Ceresium simplex**, Gyllenh. — *Stenochorus simplex*, Gyll. in Schönh. Synonymia Insect. App. I, 3, p. 178. — *Arhopalus ambiguus*, New., The Entomologist, vol. I, p. 246.

Taken at Prome in Pegu. A widely distributed species. The *Oemona philippensis* of Newman (Entomologist I, p. 246) is distinct from this species, and is to be referred to the genus *Examnes* of Pascoe.

36. Obrium posticum, sp. n. *Fulvo-testaceum; capitis fronte, pedibus (dimidiis basalibus femorum exceptis), elytrorumque apicibus piceo-nigris; antennis articulis 4 basalibus nigris, ceteris obscure fulvis. Long. 9 mm.*

Mandalay. One example. This species has the general appearance of *Obrium cantharinum,* Linn. It is of about the same size; and the punctuation is very similar in the two species. It may be distinguished by the black front of the head, the black apex of the elytra and the black legs. The prothorax appears slightly broader, in proportion to the size of the insect, than in *O. cantharinum;* and the tubercle on the posterior median part of the disk is almost quite obsolete.

Ibidionidum, g. n.

Head projecting; front declivous, short, separated from the epistome by a transverse groove. Eyes prominent, coarsely facetted. Antennae a little longer than the body in both sexes, with the scape curved and gradually thickened towards the apex, the second joint short, the third scarcely more than twice as long as the second, the fourth a little longer than the third, the fifth and following joints subequal, each much longer than the fourth. Prothorax elongate, cylindrical, with a conical tubercle at the middle of each side, constricted behind the base of the tubercles, and also, but less strongly so, in front of them. Elytra a little more than twice as long as the prothorax, gradually narrowed posteriorly, with the apices rounded. Pro- and meso-sterna simple, the former dilated behind so as to complete the closure of the anterior cotyloid cavities. Anterior and intermediate cotyloid cavities completely closed in on the outside. Legs moderately long; femora abruptly clavate at the apex, pedunculate at the base. Abdomen much narrowed towards the apex.

Male. First abdominal segment elongated, equal in length to the four succeeding segments combined.

Female. First and second abdominal segments elongated, the first a little longer than the second, the latter as long as the succeeding segments combined.

37. **Ibidionidum Corbetti**, sp. n. (Pl. I, fig. 3). *Sparsim ciliatus, testaceus; prothorace quam latitudine basis duplo longiore, supra utrinque leviter binodoso, sparsim ciliato; elytris testaceo-fulvis, griseo subtiliter pubescentibus, seriatim punctatis, punctis setigeris.* Long. 7-10 mm.

Hab. Burma. Two examples, a male and female, were captured by Mr Fea at Yenang Young; the species has also been taken at Paungdè by Mr G. Q. Corbett.

The elongated thorax, somewhat raised and curved anteriorly, gives this little species the aspect of one of the *Ibidioninae*. Its characters show, however, that it belongs to the *Obrionides*.

38. **Thranius simplex**, sp. n. *Fusco-testaceus, corpore subtus pedibusque testaceis, his clavis femorum piceis, antennis fusco-ferrugineis articulis* 8°, 9°-*que testaceo-fulvis: capite prothoraceque griseo-flavo pubescentibus, hoc antice lateraliter compresso, dorso gibboso et in gibbo summo sub-asperato: elytris fusco-ferrugineis, crebre fortiterque punctatis, utrisque linea longitudinali paullo elevata et ad basin extremam tuberculo parvo instructis, apicibus utrisque in spinam productis.* Long. 20 mm.

Carin Mts. (district of Ghecù). One example.

This species may be distinguished from *T. gibbosus* (with which it agrees somewhat closely in structural characters) by the rather dark brown colour of the entire elytra.

So far as I can judge from the description of *Singalia*, and of its type species, this genus, which Lacordaire placed in the group of the *Saperdides*, appears to me to have been founded upon the *Thranius gibbosus* of Pascoe. Lacordaire himself recognized that the genus presented characters which were quite unusual in the group. But I feel strongly convinced that he overlooked the fact that the characters were not only foreign to the group, but even to the sub-family of the *Lamiidae* itself. Lacordaire was not unacquainted with the genus *Thranius*, for he has given a very full description of it in its proper place.

Bates has also added to the genus *Singalia* a species from Japan. I have seen the type of this species (*S. rufescens*, Bates, Journ. Linn. Soc. XVIII, p. 258), which is in the collection of Mr Lewis. It belongs to the genus *Thranius*, and somewhat resembles the species described above, but is shorter and stouter in form, and not quite so dark in colour.

39. **Pyresthes birmanica**, sp. n. ♂. *Nigra, prothoracis dorso (margine antico excepto) elytrisque rufis, abdomine rufescente basi medio nigro; prothorace antice valde sub-abrupteque constricto, supra transversim haud fortiter striato, opaco, macula parva medio disci levi, polita, subtus rugoso-punctato; scutello nigro, postice acuminato: elytris crebre haud fortiter punctatis, opacis, apicibus rotundatis, sutura brevissime mucronatis, lateribus paullo pone basin leviter sinuatis: antennis corpore fere aequalibus, articulo 3° quam 1° vel 4° longiori.*

♀. *A mare differt antennis brevioribus, prothorace antice minus abrupte constricto, supra rugoso-punctato, abdomine omnino rufescente.*

Thagatà (♂) and Moulmein (♀) in Tenasserim. In the male type from Thagatà the prothorax is rather strongly and somewhat abruptly constricted at the apex; between this and the middle its sides are slightly dilated and rounded, and, between the middle and the base, feebly sinuate; the upper side is crossed by some feeble and not very regular ridges, and is without any distinct signs of punctuation. The elytra are only slightly sinuate on each side just behind the base, and not distinctly emarginate as in *P. haematica*, Pasc.; they are entirely red, with the exception of a slight infuscation on the anterior part of the suture, and are nowhere glossy, the punctures being closely and evenly distributed over the whole surface, and not much larger towards the base than at the apex. The third joint of the antennae is a little longer than the first or fourth.

The female example, taken at Moulmein, differs from the male just described in having much shorter antennae; its prothorax is less abruptly constricted at the apex, and has the upper surface very closely and somewhat rugosely punctured:

its abdomen is entirely of a dull reddish tint. The agreement in other respects is, however, so close, that I have no hesitation in referring it to the same species.

P. scapularis, Pasc., seems (so far as I can judge of it from the figure and description) to be nearly related to the present species.

40. **Pachyteria superba**, Gestro, Ann. Mus. Civ. di Genova, Ser. 2. vol. VI, p. 128.

Taken at Bhamô in Upper Burma.

41. **Chloridolum alcmene**, Thoms., Systema Cerambycidarum. Appendix, p. 568.

This species is represented by examples which may be arranged in two series according to their size, but which show no structural character by which I can distinguish them.

The larger forms were taken at Carin (Asciuii Chebà) Alt. 1200-1300 m. The smaller variety was taken at Carin (Chebà). Alt. 900-1100 m., and also at Thagatà in Tenasserim.

42. **Leontium sinense**, Hope, Proc. Ent. Soc. Lond. 1841, p. 63; Trans. Ent. Soc. Lond., IV (1843), p. 17.

One example from Carin (Chebà) Alt. 900-1100 metres.

In the catalogue of Gemminger and Harold this species is to be found placed as a synonym of *L. argentatus*, Dalm. The two species are, however, distinct. In *sinense* the joints of the antennae from the 7.th to the 10.th are each acutely angular at the inner apex; in *argentatus* the joints from the 5.th to the 10.th are each produced into a spine at the inner apex. The apex of the last ventral segment in the female of *sinense* is distinctly emarginate; the corresponding segment in *argentatus* (φ) is rounded and entire at the apex.

L. sinense is most nearly allied to *L. subtruncatum*, Bates, which it resembles by its obtusely rounded or subtruncate apices of the elytra; but from which it is to be distinguished by the structure of the prothorax.

43. **Polyzonus flavocinctus**, sp. n. ♂. *Viridi-metallicus; prothorace transversim strigoso, strigis medio dorsi intricato-confluentibus; pronoto antice in medio paullo rotundato-producto; elytris fasciis*

duabus angustis flavis, atro-violaceo limbatis, ornatis; corpore subtus argenteo-sericeo, segmento quinto abdominis leviter sinuato, fere transverso-truncato, sexto sinuato-emarginato.

A single male specimen was taken at Carin Mts. (Chebà district) at an altitude between 900 and 1100 m. The species occurs also in Tenasserim.

This species closely resembles *P. bizonatus*, White; but differs from it as follows: Prothorax more regularly strigose towards the sides of the disk; the anterior margin of the pronotum somewhat obtusely produced in the middle. Elytra with a glabrous rugose-punctate area, somewhat cordate in shape, just behind the scutellum. (The corresponding area in *P. bizonatus* is bluish-black and opaque and not more strongly punctate than the rest of the surface). The posterior margin of the fifth abdominal segment in the male is almost directly transverse, that of the sixth segment is sinuately emarginate. (In *P. bizonatus* both segments are deeply and sinuately emarginate behind).

The present species does not differ from *P. tetraspilotus*, Hope by any good structural characters, and may possibly prove to be only a variety, distinguished by having two transverse bands, instead of four spots, on the elytra.

44. **Polyzonus bizonatus**, White, Cat. Longic. Brit. Mus., p. 171.
Rangoon. One female example.

45. **Anubis inermis**, White, Cat. Longic. Brit. Mus., p. 171.
This species was taken at Carin Mts. (district of Chebà), at Kawkareet in Tenasserim and also at Rangoon.

46. **Anubis fimbriatus**, Bates, Cistula Entomologica, II, p. 412.
Carin Mts. (district of Chebà).

47. **Anubis bipustulatus**, Thoms., Syst. Ceramb. App., p. 569.
Carin Mts. (district of Chebà), and at Teinzò and Bhamò in Upper Burma.

48. Anubis rostratus, Bates, Cist. Ent. II, p. 412.
Carin Mts. (district of Chebà). One example.

49. **Rosalia decempunctata**, Westw. — *Purpuricenus decempunctatus*, Westw., Cab. of Orient. Entom., p. 59, pl. 29, fig. 2
— *Eurybatus decempunctatus*, White, Longic. Brit. Mus. p. 141;

Pascoe, Longic. Malay., p. 597. — *Rosalia decempunctata*, Lameere, Ann. Soc. Ent. Belg. XXXI, p. 163, t. 3, fig. 6 (♀).

Mt. Mooleyit in Tenasserim; alt. 600-1700 metres.

A variety of this species, in which there are no spots to the elytra, was taken at Carin Mts. (district of Ghecù), alt. 1300-1400 m. The species occurs also in N. India (Sylhet, Darjeeling, Sikkim) and in Borneo and Java.

50. **Rosalia formosa**, Saund.

Catcin Mts. 1886.

51. **Xylotrechus Hampsoni**, Gahan, Ann. & Mag. Nat. Hist., Ser. 6, vol. V, p. 54, pl. VII, fig. 1.

Carin Mts. (district of Chebà); alt. 900-1100 metres. One example.

This species was described from a single specimen obtained by M.' Hampson in the Nilghiri Hills, South India. It is one of those intermediate forms that connect the genera *Xylotrechus* and *Clytus*. I have placed it in the former genus, owing to the presence on the front of the head of a median longitudinal groove with raised edges in the form of carinae, and of a short sharp edge or carina on each side just over the antennary condyle.

52. **Xylotrechus Gestroi**, sp. n. (*Pl.* I, *fig.* 4). *Capite prothoraceque flavo-pubescentibus; prothoracis dorso signo cruciformi fusco cujus brachiis macula triangulari utrinque sublatis; lateribus utrisque macula parva rotunda fusca; elytris fusco-velutinis, fasciis flavis, una prope basin literae x parum simili, una submedia triangulari, tertia ad apicem transversa, ornatis; antennis pedibusque testaceis, femoribus posticis apice infuscatis. Long.* 11. *Lat.* 3 *mm.*

One example taken at Shwegoo on the Upper Irrawaddy.

Head and prothorax with a yellowish pubescence. The front of the head with a median black line; obsoletely grooved between the lower lobes of the eyes, and with a short sharp edge on each side just over the antennary condyle. Prothorax rounded at the sides, marked above by a dark brown figure somewhat resembling a cross whose arms are supported on each side by a triangular or nearly oblong spot; on each side of the prothorax there is a small rounded fuscous spot. Scutellum yellow. The elytra may be described as dark velvety brown, with yellowish bands, of which the first, near the base, is somewhat roughly *x* shaped; the second triangular, with its apex just in front of- its base behind- the middle, and its sides slightly concave; the third is a transverse fascia occupying the apical fifth but without reaching the outer margin. That part of the elytra between the basal margin and the *x* shaped patch is testaceous in colour and is partly clothed with yellowish pubescence. The apices are truncate, each very shortly mucronate at the outer angle. Body underneath with a yellowish pubescence; each of the last four abdominal segments dark brown and nitid anteriorly. Legs and antennae testaceous, with the hind femora infuscate towards the apex.

The peculiar marking of this pretty little species will admit of its being easily recognized. I have named it in honour of D.r Gestro who has already done so much to advance the study of Coleoptera.

53. **Xylotrechus quadripes**, Chevr., Mém. Soc. Roy. des Sciences de Liège, vol. XVIII (1863), p. 315; — Dunning, Trans. Ent. Soc. Lond. 1868, p. 126 and fig. — *Cucujus coffeophagus*, Richter, Proc. Agri-Hort. Soc. Madras 1867.

Carin Mts. (Chebà): alt. 900-1100 m. Two females. Palon in Pegu one male. The species occurs also in India — Sylhet and Madras — and in Siam; and is of economic importance, inasmuch as it is injurious to the coffee-plant (vide Dunning, op. supra cit.). Chevrolat described the female. In this sex the front of the head has five nearly parallel and equidistant raised lines or carinae, of which the two intermediate are shorter than the rest and do not extend to the vertex. In the male the front of the head has two elongate, very finely granular and opaque black spots, one on each side between the median and the outer raised lines, and each is bounded by a very narrow and slightly raised line. In other respects, a slight difference in the length of the antennae excepted, the two sexes closely resemble each other.

54. **Xylotrechus Phidias**, Newm. — *? Clytus Buqueti*, Lap. & Gory, Monograph. des Clytus (1841), p. 86, pl. 16, fig. 99. — *Clytus Phidias*, Newm., The Entomologist, vol. I (1842) p. 246 ($\male$).

Two examples ($\male$ and $\female$) taken at Thagatà in Tenasserim. Male with a thin greyish pubescence on the head and prothorax, with a faint sub-nude blackish patch on the middle of the prothorax posteriorly. Head and prothorax of the female with a yellowish pubescence; with a distinct black band extending along the middle of the pronotum almost from the base to the apex, and widening out posteriorly; with a rounded black area or spot on each side of the prothorax. The bands of the elytra in both sexes are as represented in the fig. of *X. Buqueti*, L. & G. (loc. supra cit.) The female differs further from the male by the form of the last dorsal segment of the abdomen; this segment is strongly emarginate at the apex in the female, and is rounded and not emarginate in the male. The single male specimen in the present collection agrees well with the male type of *X. Phidias*, Newm. The female resembles closely the figure of *X. Buqueti*, L. & G. in all but the colour of the pubescence on the head and prothorax which in *Buqueti* is described, and also shown in the plate, to be cinereous grey and not yellowish. The black spots on the sides of the prothorax are also larger than in *Buqueti*; though two female examples from Burma in the British Museum have these spots smaller than in the female taken by Signor Fea. It is possible that *X. Buqueti* has been described from the male of a very closely allied species, and it is with doubt, therefore, that I regard it as synonymous with *X. Phidias*.

55. **Xylotrechus vicinus**, Lap. & Gory. — *Clytus vicinus*, Lap. & Gory, Monograph., p. 38, pl. 8, fig. 47.

Taken at Rangoon; at Meetan and Kawkareet in Tenasserim. It occurs also at Moulmein in Burma, and at Calcutta and Bhotan in N. India.

The males differ from the females by the same characters of the last dorsal segment as in *X. Phidias*, Newm.

56. **Perissus proprius**, sp. n. ($\male$) *Niger; capite prothoraceque*

leviter griseo-pubescentibus; prothorace supra sparsim asperato-punctato, medio plaga transversa nigra, lateraliter rotundato, pone medium latiore, deinde usque ad basin valde angustato; scutello nigro, cinereo-marginato; elytris fusco-nigris, fasciis cinereis- una ad basin, secunda ante medium valde contorta, tertia vix pone medium transversa et ad suturam angulato-dilatata, quarta ad apicem sub-obliqua; apicibus sub-obliquiter truncatis, angulis breviter dentatis; antennis medium elytrorum paullo excedentibus.

(♀) *Mare differt fasciis prothoracis elytrorumque nonnihil flavescentibus, antennis medium elytrorum nec attingentibus.* Long. (♂ ♀) 6-12 $^1/_2$ mm.

Thagatà in Tenasserim; and Mandalay in Upper Burma.

Prothorax rounded at the sides, somewhat widened from the base to beyond the middle and thence strongly narrowing to the apex; somewhat sparsely asperately punctured along the middle above, the disk with two black spots or a transverse black patch. The first cinereous band of the elytra runs across the base on each side from the scutellum to the humeral depression where it is slightly turned backwards; the second begins at the suture a little behind the scutellum, diverges from the suture while passing backwards for some distance, then curves round to take an anterior and outward direction, and is again sharply bent backwards just before reaching the outer margin: the third band, placed just behind the middle, is narrow and slightly oblique in its outer half, and is angulately dilated both anteriorly and posteriorly at the suture; the fourth is a rather broad and somewhat oblique band placed at the apex. The body underneath is closely covered with an ashy-white pubescence. The legs are black, with a faint greyish pubescence; the first joint of the posterior tarsi is nearly three times as long as the two succeeding joints combined.

The female differs from the male, not only by its shorter antennae, but also by the yellowish-grey colour of its elytral bands, and by having an illdefined longitudinal yellowish-grey band on each side of the prothorax above.

This species seems closely allied to *P. cinereus*, L. & G. (Mon.

des Clytus, p. 68, pl. 13, fig. 79), and to *P. fuliginosus*, Chevr.
— the type of Chevrolat's genus *Amauresthes*. It has a very similar style of marking to the former, but differs from both by the sparser punctuation along the middle of the pronotum, and by the black patch on the middle of the disk. I do not find that *Amauresthes* can be distinguished by any definite characters from *Perissus*. The three species just mentioned agree in having their prothorax a little bulged out just behind the middle; but, so far as I can see, there is no other character of the least generic importance by which they differ from typical species of *Perissus*.

The *Clytus cinereus* of Lap. & Gory (= *C. Duponti* (Dej.) Muls.) appears under the genus *Clytus* in Reitter's Catalogue of European Coleoptera: but if the genus *Perissus* is to be retained, as, for convenience sake, I think it may, it ought to include *C. cinereus*.

57. **Perissus persimilis**, sp. n. *Precedenti similis, sed prothoracis lateribus magis parallelis, discoque sine plaga transversa nigra, scutello omnino cinereo. Elytrorum fasciis ante-mediis minus contortis.* Long. 8 mm.

Two examples taken at Thagatà in Tenasserim.

This species has at first sight a strong resemblance to the preceding, but may be distinguished by its more parallel-sided prothorax, which is moreover without a transverse black space above. The scutellum is wholly cinereous. The antemedian band of the elytra begins a little behind the scutellum, passes backwards and outwards, diverging rather strongly from the suture, is then turned slightly forward, and reaches the outer margin without undergoing a second bend.

58. **Perissus mutabilis**, sp. n. *Capite prothoraceque rufo-ferrugineis aut nigris, minute granulosis, griseo subtilissime pubescentibus; prothoracis dorso plaga transversa sub-nuda, margine basali utrinque albescente; scutello albo; elytris nigro-opacis, fasciis griseis et cinereis ornatis, abdomine et fasciis pectoris cinereis; pedibus antennisque nigris, his griseo-pubescentibus.* Long. 8-14 mm.

Thagatà in Tenasserim.

Head and prothorax minutely granulose, their colour varying from ferruginous red to black. The prothorax with a transverse sub-nude space on the middle of the disk. Scutellum and the latero-basal margins of the prothorax whitish. Elytra dull velvety black with ashy and grey markings, of which the first consists of a band on each which commences at the suture just behind the scutellum, passes obliquely backwards for some distance without strongly diverging from its fellow, and is then directed almost straight outwards so as nearly to reach the external margin, the second is a transverse or slightly arcuate band crossing the elytra at about the middle, and gradually becoming wider towards the suture so as to have a somewhat shortly triangular form; the third is a transverse band, of a somewhat darker grey colour, which occupies a little more than the apical fifth. The first two segments of the abdomen, and two bands on each side of the breast (one longitudinal and one oblique) are of an ashy-white colour; the hinder segments of the abdomen have a darker grey pubescence.

The head in this species is more flattened between the antennae and more regularly rounded from the vertex to the front than in other species of the genus. In this respect it approaches the African genus *Culanthemis*, Thoms.

For the four following species I use the generic term *Caloclytus* (¹) in preference to *Clytanthus*, believing that the latter name would be better restricted to certain Central American species which are distinguished by having a longitudinal groove or depression on each side of the elytral suture limited on the outside by a more or less distinct carina. The genus *Clytanthus*, thus restricted, is more allied to *Ochresthes*, than to the European and Asiatic species hitherto included in the former genus.

59. Caloclytus **14-maculatus**, Chevr. — *Anthoboscus 14-maculatus*, Chevr., Mém. Soc. Roy. des Sc. de Liége, vol. XVIII (1863), p. 295.

Thagatá in Tenasserim. One example.

(¹) *Caloclytus*, Fairm. (type *C. semipunctatus*, Fabr.)

The type specimens and other examples in the Brit. Mus. collection, are all from the Nilghiri Hills, South India.

60. **Caloclytus annulatus**, Hope. — *Clytus annulatus*, Hope, Gray's Zool. Misc. (1831), p. 28.

Taken at Thagatà in Tenasserim, and at Carin (Asciuii Ghecù). Alt. 1400-1500 m.; occurs also in N. India (Nepal).

Head and prothorax with a rather close ashy-grey pubescence. The prothorax with three black spots, one larger medio-dorsal, and one smaller, towards each side. Elytra black with greyish markings — an x-shaped figure extending from the base to near the middle, and inclosing an oval spot on each side; a transverse band, dilated at the suture, placed behind the middle; and an oblique fascia on each at the apex; apices of elytra obliquely truncate; outer angles acute, sub-spinose. Body underneath, legs and antennae with a greyish pubescence. This species comes near *C. macaonensis*, Chevr.

61. **Caloclytus ludens**, sp. n. (*Pl.* 1, *fig.* 5). *Niger; capite prothoraceque pube grisea subtiliter obtectis; prothorace elongato, cylindrico, lateraliter leviter rotundato, margine basali albo-cinereo-pubescente; scutello cinereo; elytris nigris, fasciis cinereis ornatis- una basali transversa; secunda ab sutura retro obliquiter directa, et postice extus curvata; tertia transversa, breviter triangulari et vix pone medium obsita; quarta lata transversa ad apicem; apicibus elytrorum recte truncatis, extus breviter spinosis. Long.* 11 *mm.*

Thagatà in Tenasserim. One female example.

With a thin greyish pubescence on the head and prothorax: the latter somewhat asperate above, its basal margin whitish. Scutellum grey. Elytra black, with dull-cinereous markings, consisting of (1) a transverse basal fascia; (2) a narrow band on each beginning at the suture just behind the scutellum, diverging from the suture as it passes back for a short distance, and then curved so as to run outwards and slightly forwards without, however, reaching the external margin; (3) a transverse band placed just behind the middle and having the form of a short triangle with its apex placed anteriorly at the suture; (4) a rather broad transverse band at the apex. Body underneath with a greyish pubescence, passing to whitish on the sides of the breast and abdomen. The antennae of the female are piceous, with a greyish covering, and reach to about the beginning of the posterior fourth of the elytra.

The style of marking of this species somewhat resembles that of *Perissus mutabilis;* but the species may be easily distinguished from the latter by the greater length of its antennae, and, especially, by the form of the head, which in *Perissus* is broad and level between the antennae, but in *Caloclytus* is narrower and concave in this region.

62. **Caloclytus signaticollis**, Lap. & Gory, *var.* — *Clytus signaticollis*, L. & G., Mon. du Genre Clytus, p. 103, pl. 19, fig. 122. — *Clytus oppositus*, Chevr. Mon., p. 304.

Carin Mts. (district of Chebà).

The examples taken at Carin have a narrower form, and narrower black bands to the elytra than in the specimen figured by Gory. The median spot of the prothorax is slightly transverse. This variety occurs also in Assam.

63. **Chlorophorus annularis**, Fabr. — *Callidium annulare*, Fabr., Mant. Insec., I, p. 156.

Teinzò, Bhamò and Shwegoo in Upper Burma; Carin Mts., Chebà district; Thagatà and Plapoo on Mt. Mooleyit and Malewoon in Tenasserim.

This species appears to be very widely distributed. The British Museum collection contains examples from North India, South Japan, Hong Kong, Formosa, Siam, Sumatra, Borneo and Celebes, and one example from Port Moresby in New Guinea.

64. **Demonax semiluctuosus**, White. — *Clytus semiluctuosus*, White, Longic. Brit. Mus., p. 283. — *Acrocyrta semiluctuosa*, Chevr. Mém. Soc. Roy. Sc. Liége, XVIII, p. 260. — ? *Rhaphuma praecana*, Chevr., l. c., p. 277. — *Clytanthus Mouhoti*, Pasc., Trans. Ent. Soc. Lond., Ser. 3, vol. III, p. 604.

Taken at Thagatà and Meetan in Tenasserim; and at Carin Mts. (Chebà district), alt. 900-1100 m. The species occurs also at Moulmein; and in India and Malacca.

65. **Demonax dignus**, sp. n. *Cupite nigro, griseo-pubescente; pro-*

thorace elongato, ferrugineo, opaco, minute granuloso, margine antico nigro, margine basali albo-pubescente; scutello albo; elytris nigro-fuscis, macula utrinque antice, fascia ad medium obliqua, antice ad suturam prolongata, et fascia transversa ante apicem, albo-cinereis; segmentis duobus abdominis primis et maculis thoracis albescentibus, segmentis 3 posterioribus abdominis griseis; pedibus antennisque nigris, griseo subtiliter pubescentibus, antennis (♂) quam corpore vix longioribus, articulis 3° et 4° apice extus breviter spinosis. Long. 12 mm.

Two examples taken at Carin Mts. (district of Chebà). Alt. 900-1100 m.

Prothorax elongate, cylindrical, very slightly rounded at the sides, dull ferruginous-red with the anterior margin black, the basal margin whitish. Scutellum white. Elytra dark brown, with ashy white markings; consisting of a small transverse spot on each, close to the suture, at about the anterior fifth: a narrow oblique band on each at the middle which runs forward along the suture almost up to the anterior spot, and a narrow transverse band placed at some distance before the apex. Between the last band and the apex the elytra have a dark grey pubescence. Apices of the elytra transversely truncate.

This species is closely allied to *D. semiluctuosus*, White, from which it differs, inter alia, by the colour of the prothorax, and by having the median band of the elytra more oblique.

66. **Demonax literatus**, sp. n. (*Pl.* I, *fig.* 6). *Elongatus, luteogriseo dense pubescens, nigro-ornatus; prothorace quam latitudine longiori, lateribus leviter rotundatis, supra maculis quatuor nigris antice transversim dispositis, et macula longitudinali subnigra medio prope basin. Elytris luteo-griseo-pubescentibus, utrisque vittis maculisque nigris- duabus inter basin mediumque V-formantibus, macula lunata vix pone medium cujus cornubus antice prolongatis, et macula sub-rotundata ante apicem; apicibus truncatis angulis breviter spinosis: corpore subtus albo-cinereo-pubescenti; femoribus piceonigris, tibiis antennisque testaceis, his (♂) quam corpore longioribus, articulis 3°-6ᵐ apice extus breviter spinosis. Long.* $9^1/_2$-11. *Lat.* $2^1/_4$ *mm.*

Two examples from Carin Mts. (Chebá district). Alt. 900-1100 m.

With a buff grey pubescence above. Prothorax with four black spots placed transversely in front of the middle, of which two only are visible when looked at from above, with one, or three, narrow and less distinct black spots, running longitudinally between the middle and the base. Elytra each with a V-shaped black figure between the middle and the base, the inner branch of which is joined to its fellow of the opposite side by a short transverse bar just behind the scutellum, so that the combined figures are somewhat in the shape of a W; with a crescentic black spot or fascia behind the middle, the horns of which are prolonged anteriorly, the inner horn joining, at the suture, the corresponding one of the other elytron; with a somewhat rounded black spot about midway between the crescentic spot and the apex.

This species has (notwithstanding the difference in markings) a general resemblance to *Grammographus Horsfieldi*, White, and I should have included it in the latter genus, were it not that certain joints of the antennae are shortly spined — a character wanting to *G. Horsfieldi*.

67. **Demonax reticollis**, sp. n. (*Pl.* I, *fig.* 7). *Niger; capite, prothorace, pedibus antennisque et corpore subtus cinereo-griseo-pubescentibus; prothorace oblongo-ovato, supra lateraliterque minute reticulato-corrugato, basi utrinque albo-marginato: scutello albo; elytris fuscis (antice pallidioribus), fasciis albo-cinereis ornatis; apicibus truncatis, extus breviter dentatis; antennis (♀) quam corpore brevioribus, articulis 3°, 4°-que apice extus longe spinosis. Long.* 13 *mm.*

One female example taken at Carin Mts. (Chebá district) at an alt. of 900-1100 m.

Head, prothorax, legs, antennae and underside of the body with a greyish pubescence; the basal margin of the prothorax on each side, and the sides of the breast and two first abdominal segments whitish. Prothorax slightly rounded at the sides, covered above and at the sides with thin feebly raised lines which run together to form a very fine reticulation. Elytra dark brown, somewhat lighter in colour towards the base, marked with ashy-white bands, of which the first begins at the scutellum, accompanies the suture for a short distance, then slightly diverges and, at about the termination of the anterior third of the elytra, is turned straight outwards towards the margin which it does not quite reach, the second band is placed transversely a little behind the middle and is somewhat produced anteriorly along the suture, the third band, also transverse, is at the apex. The antennae in the female (the only sex represented) are somewhat shorter than the body, with a rather long and slender spine at the outer termination of the third and fourth joints.

68. **Demonax macilentus**, Chevr. — *Aerocyrta macilenta*, Chevr. Rev. et Mag. Zool. X (1852), p. 82 ; Mém. Soc. R. Sc. de Liège XVIII, p. 260.

Carin (district of Chebà), alt. 900-1100 m.

The specimens taken at Chebà differ slightly from the type as follows : — The two black spots of the prothorax are more distinct. The second ashy-grey band of the elytra is not in the form of a short broad W as in the type, but is interrupted at the suture so as to have the appearance of a distinct and separate V on each elytron.

69. **Epipedocera zona**, Chevr., Mém. Soc. Roy. Liège, XVIII, p. 340.

Two example, one taken at Teinzò, the other at Bhamò in Upper Burma. In the Bhamò specimen the femora are black. In normal forms of the species the femora are reddish. The species occurs also in N. India (Darjeeling and Nepal).

70. **Nida flavovittata**, Pasc., Ann. & Mag. Nat. Hist., Ser. 3, vol. XIX, p. 312.

Carin Mts. (district of Chebà) ; occurs also in Pegu.

71. **Polyphida Feae**, sp. n. (*Pl.* I, *fig.* 8). *Capite prothoraceque griseo-aurato-pubescentibus; elytris fortiter punctatis, aurato-pubescentibus, utrisque fasciis glabris chalybeatis — duabus transversis prope medium, tertia longitudinali ab humero versus medium extensa; apicibus truncatis, angulis breviter dentatis; corpore subtus sericeo-pubescente pedibus griseo tenuiter pubescentibus. Long.* 12. *Lat.* 2.75 *mm.*

Mount Mooleyit in Tenasserim.

Head with a somewhat golden pubescence; with a median glabrous impressed black line extending from the epistome to the occiput. Antennae a little longer than the body; joints 1ˢᵗ and 4.ᵗʰ subequal, each scarcely more than half as long as the third; fifth and following sub-equal, each longer than fourth and shorter than third. Prothorax with its length to breadth in ratio of 3 to 2; somewhat rounded at the sides; slightly constricted at the base, narrowed from the middle up to the anterior margin, with a very feeble constriction about midway between these two points; covered on all sides with a somewhat golden pubescence which is rather duller on the upper side. Elytra strongly punctured; chalybeate-blue, with a golden pubescence which more or less closely covers the upper surface with the exception of two transverse and slightly oblique bands on each elytron near the middle, and a less clearly defined longitudinal band passing from the shoulder to join the anterior of the two transverse bands.

The arrangement of the pubescence in this species somewhat resembles that of *Glaucytes interrupta*, Oliv. This character enables it to be easily distinguished from *Polyphida clytoides*, Pasc., in which the elytra have a silvery pubescence arranged in a pattern similar to that of many *Clytinae*.

72. **Euryphagus Lundi**, Fabr. — *Cerambyx Lundii*, Fabr., Entom. Syst. II, p. 247.

Carin Mt. (district of Cheba).

73. **Philagathes sanguinolentus**, Oliv. — *Cerambyx sanguinolentus*, Oliv., Entomologie IV, 67, p. 93, pl. 20, fig. 155.

Varieties of this species were taken at Thagata in Tenasserim and at Bhamô and Teinzô in Upper Burma. Similar varieties occur also in N. India.

74. **Nericonia nigra**, sp. n. *Nigro-opaca, femorum basi albo-flava, capite prothorace antennarumque scapo creberrime punctulatis; prothoracis dorso medio sub-rugosulo; elytris fortiter punctatis, punctis sub-transversis et in seriebus longitudinalibus inter lineas elevatas ordinatis. Long. 8 mm.*

Carin Mts. (Gheeu-district) alt. 1300-1400 m.

Black, with the exception of the bases of the femora which

are yellowish-white. Furnished with some long, rather sparsely scattered sub-erect setae. Upper part of the head, the prothorax and the scape of the antennae minutely and very closely punctured, with the punctures on the middle of the pronotum running together in a longitudinal direction and imparting to the disk a slightly rugose appearance. Elytra with strong, somewhat transversely elongated, punctures arranged in four rows on each. The inner-most row is close to the sutural edge, and is made up of smaller punctures than those of the other rows; the second runs along the middle of the dorsal part of the elytron and has on its inner side, near the base, a feebly raised line; the third and most distinct row sets out from the basal depression just above the shoulder, and, along its inner border, has a clearly marked raised line; the fourth row runs back from the humeral prominence, and its punctures are slightly distant from one another. The raised lines and the large punctures do not extend on to the posterior sixth of the elytra; but this part, and the interstices over the remainder of the elytra are very minutely and closely punctulate, and covered by a very faint dark-brown or blackish pubescence. The apices of the elytra are rounded. The legs, except at the base, are blackish, sub-nitid and sparsely setose. The antennae (♀?) are nearly twice as long as the body, they are covered with a very short pubescence, and the joints are furnished underneath with some very long close-lying hairs; the third joint is about equal in length to the fourth.

Though readily distinguishable from *Nericonia trifasciata*, Pasc. the above species agrees with it very well in structural characters. The two species placed by Pascoe in *Melegena* differ by no very strongly marked characters, but may be distinguished by the emarginated and toothed (or spinose) apices of the elytra; and their somewhat less strongly clavate femora. In all of there as in the other members of the group *Disteniides* there is an oblique groove or notch on the tibiae of the middle pair; and there is also an oblique, though less distinct, groove on the ventral side of the tibiae of the anterior pair similar to that which is so characteristic of the Longicorns of the subfamily Lamiidae.

LAMIIDAE

75. **Arctolamia villosa**, Gestro, Ann. Mus. Civ. di Genova, Ser. 2.ª, vol. VI, p. 129; op. cit., Ser. 2.ª, vol. X (XXX), p. 222 and fig. Taken at the Catcin Cauri Mts. (= Kakhien Hills of English Maps).

76. **Arctolamia fasciata**, Gestro, l. c., Ser. 2.ª, vol. X (XXX), p. 221 and fig.

This fine species was taken in the valleys of Sittang and Salween (Carin Mts.); and also in the district of Chebà at an altitude of 900-1100 m.

77. **Stratioceros princeps**, Lacord., Genera des Coléoptères, vol. IX, Pt. 1, p. 303; Brongn., N. Archiv. Mus. (3) III, pl. X, fig. 10.
— *Cacoscapus Mouhoti*, Thoms., Rev. et Mag. de Zool. 1878, p. 47.

Taken at Carin Mts. (district of Ascinii Chebà) at an altitude of 1200-1300 m.; occurs also in Siam.

78. **Lamiomimus officinator**, White. — *Monohammus officinator*, White, Proc. Zool. Soc. 1858, p. 109. — *Archidice quadrinotata*, Thoms., Systema Cerambycidarum, p. 554.

One example taken at Catcin Cauri Mts. (= Kakhien Hills); occurs also in N. India.

79. **Leprodera bisignata**, sp. n. *Cinereo fulvoque et brunneo pubescens, capite linea media nigra et supra vittis brevibus duabus nigris, tuberculis antenniferis antice punctatis; scapo antennarum asperato-rugoso, articulo 3° basi sub-asperato; prothorace lateraliter acute spinoso, supra vittis quatuor indistinctis fulvis: elytris punctatis, utrisque basi maculis duabus arcuatis nigrovelutinis, granulis paucis nitidis includentibus, humeris nigro-granulosis; apicibus rotundatis, pedibus anticis (♂) elongatis, tibiis ejusdem ante apicem tuberculatis; antennis (♂) quam corpore duplo longioribus. Long.* 24 mm.

One example taken at Catcin Cauri in Upper Burma.

Head with a pubescence partly ashy-grey and partly fulvous, with two short black vittae above extending from the thorax almost as far as the upper lobes of the eyes and gradually narrowing in front. Prothorax above with four indistinct fulvous vittae, two near the middle and one on each side just above the lateral spine. Scutellum pubescent, with a median line naked black and glossy. Elytra punctate, with an ashy-grey pubescence, interrupted by faint irregular brownish patches; each elytron at the base with two short arcuate velvety-black vittae which include some black shining granules, the shoulders also are granulate. The body underneath and the legs have an ashy or fulvous pubescence, which is interrupted by numerous small naked points. The mesosternum is very feebly carinate in the middle. Antennae with a greyish pubescence.

This species may be easily recognized by the peculiar eye-like arrangement of spots at the base of each elytron. There is a similar eye-like spot in *Morimus? diophthalmus*, Pascoe but the latter species is easily to be distinguished by its shorter, more robust, and strongly punctured antennae; by its shorter metasternum; and by many other characters.

80. **Leprodera stephana**, White. — *Monohammus stephanus*, White, Proc. Zool. Soc. 1858, p. 406.

Taken at Carin Mts. (Chebà district) at an altitude of 900-1100 m.; occurs also in N. India.

81. **Leprodera insidiosa**, Gahan, Ann. & Mag. N. Hist., Ser. 6, vol. II (1888), p. 391.

Taken at Bhamó in Upper Burma, and at Thagatà in Tenasserim. The species is found also in North India.

82. **Epepeotes luscus**, Fabr. — *Lamia lusca*, Fabr., Mant. Insect. I, p. 139.

Taken at Thagatà in Tenasserim and at Palon in Pegu. This is a widely distributed species, occuring also in Siam, Malacca, Borneo, Java, Sumatra and Nias Isl.

83. **Epepeotes uncinatus**, Gahan, Ann. & Mag. N. Hist., Ser. 6, vol. I (1888), p. 271, pl. XVI, fig. 2.

Taken at Carin Mts. (Chebà district); appears to be common in N. India (Assam, Sikkim etc.).

84. **Epepeotes vestigialis**, Pasc.? Longic. Malay., p. 301.

Thagatá in Tenasserim. One example.

85. **Pelargoderus antennatus**, sp. n. *Fuscus, tenuissime griseopubescens; elytris maculis parvis fulvis irregulariter dispersis, prothorace quam latitudine paullo longiori, antice paullo angustato; supra leviter rugoso, margine antico leviter arcuato-emarginato, elytris basi granulosis deinde punctatis, apicibus rotundatis, corpore subtus griseo fulvoque sparsim pubescente; pedibus griseis, femoribus punctis numerosis glabris nigris; antennis ($\male$) quam corpore plus duplo longioribus, articulis primo tertioque scabroso-punctatis, tertio quam primo fere triplo longiori, versus basin incrassato. Long.* 22 mm.

Carin Mts. (district of Chebá), alt. 900-1100 m.

Dark brown with a very faint greyish pubescence; prothorax slightly and almost obsoletely rugose above, the anterior margin of the pronotum slightly bowed inwards : the disk with two faint fulvous spots, and the sides each with an indistinct fulvous vitta. Elytra with numerous small black granules at the base, beyond which they are somewhat thickly punctured with the punctures diminishing in size towards the apex; covered with a very faint greyish pubescence, which is interrupted by a number of irregularly scattered small fulvous spots. Mesosternal process with a small conical tubercle. Anterior legs elongated, their tibiae each with a tooth placed a short distance before the extremity.

86. **Monohammus ocellatus**, Gahan, Ann. & Mag. N. Hist., Ser. 6, vol. II (1888), p. 262.

Taken at Teinzò and Bhamò in Up. Burma, at Thagatá in Tenasserim, and at Carin Mts. (Chebá district).

87. **Monohammus bimaculatus**, Gahan, l. c., p. 260.

Taken at Meetan in Tenasserim, and at Bhamò in Up. Burma, occurs also in Siam and in North India.

88. **Monohammus versteegi**, Ritsema, Notes Leyden Mus., vol. III (1881), p. 51; Midden Sumatra Nat. Hist. Fauna I. p. 133, pl. 3, fig. 4.

Thagatá in Tenasserim.

This species, which was founded upon a specimen from the mountains of Sumatra, appears to be common in the hilly districts of North India. I have seen many examples from Assam, and a large series from Mungphu in Sikkim. In the Sumatran example, figured by Ritsema, the elytral spots seem to be of much less than the average size. The few examples from Burma that I have seen may be regarded as constituting a variety, characterized by the distinct, though by no means strong, tuberculation of the mesosternum. The elytral spots in this variety form seven irregular transverse series. I have noticed a tendency towards a tuberculation of the mesosternum in some of the North Indian examples; so that I am inclined to believe that we have here a single variable species, rather than two or three distinct, though closely allied, forms.

89. **Monohammus dubius**, sp. n. *Fuscus, prothoracis dorso basi vittisque duabus longitudinalibus utrinque fulvis, elytris maculis parvis, irregulariter dispersis, fulvis, maculis versus apicem confluentibus: capite inter tuberculos antenniferos anguste triangulariterque sed non profunde concavo; prothorace utrinque haud fortiter tuberculato; elytris dense punctulatis, apicibus rotundatis; antennis (σ) quam corpore triplo longioribus; scapo brevi, obconico, apice valde cicatricoso ; articulo $3°$ quam $1°$ duplo longiori. Long. 15. Lat. 5 mm.*

Carin Mts. (Chebà district); alt. 900-1100 m.

Head with a rather narrow triangular cleft between the antennal tubercles; the front sub-rectangular, and almost perfectly flat; the lower lobes of the eyes rather small, not reaching more than half way to the base of the mandibles. Prothorax with a small conical tubercle, having a broad base, at the middle of each side; dark brown, with the base and two longitudinal vittae on each side above fulvous-pubescent; with a similar fulvous vitta placed very low down on each side. Elytra thickly but not strongly punctured; dark brown, with numerous small pubescent spots of a fulvous colour, which are irregularly scattered, and, in places (especially towards the apex), have a tendency to run together, thus forming small patches. Body underneath with a faint greyish pubescence, with the hind margins of the abdominal segments and the lateral pieces of the thorax somewhat fulvous. Legs ferruginous red, with the tarsi and apices of the tibiae fuscous.

90. Haplohammus cervinus, Hope. — *Monochamus cervinus,* Hope in Gray's Zoological Miscellany (1831), p. 27. — *Monohammus fraudator,* Bates, Ann. & Mag. Nat. Hist., Ser. 4, vol. XII, p. 309. — *Haplohammus fraudator,* Bates, Linn. Soc. Journ., vol. XVIII, p. 240.

Carin Mts. (Chebà district), alt. 900-1100 m.; Teinzò in Upper Burma.

This species is known also from Japan and China; and from Assam, Nepal and other districts in North India.

Hope's diagnosis of *Lamia cervina* is too brief to be of the least use in the identification of the species; especially as there are many closely allied and very similar forms. I regret, therefore, that, in deference to the law of priority, I am obliged to displace the more recent and better known name. In the British Museum collection there are two abnormal examples of this species, from Hong Kong, in which the antennae are nine-jointed.

91. Haplohammus punctifrons, sp. n. H. cervino *persimilis sed differt capitis fronte verticeque sparsim punctulatis, antennarum scapo parum graciliori, vix longiori. Long.* 15-18 *mm.*

Tenasserim: Kawkaret and Meetan; Carin (district of Chebà), alt. 900-1100 m.

Closely allied to *H. cervinus,* Hope, from which it differs by having the scape of the antennae slightly narrower and more cylindrical, and scarcely perceptibly longer; and the head sparsely punctured in front, and on that part of the vertex which lies between the antennae and the upper lobes of the eyes. In *H. cervinus* the head is usually entirely impunctate, though in three or four, out of the many examples that I have seen, there are a few widely dispersed punctures on the front.

I have seen only females of the present species; the male will probably be found to have joints $3.^{rd}$-$5.^{th}$ of the antennae thickened as in *H. cervinus.*

92. **Haplohammus longiscapus**, sp. n. *Pube grisea, sub-sericea, sat dense obtectus; capitis fronte sparsim punctulata; prothoracis dorso sparsim punctulato; elytris sat dense punctulatis, apicibus rotundatis; antennis (♂) quam corpore fere triplo longioribus, scapo elongato, usque ad apicem gradatim haud fortiterque incrassato; articulis $3°$-$5°$ paullo incrassatis. Long. 18-20 mm.*

Palon in Pegu: also taken in Tharawaddy by Mr. Corbett.

This species closely resembles *H. cervinus*, Hope, but is to be distinguished by the length of the antennal scape, which is rather more than two-thirds of the length of the third joint. When the antennae are turned back over the sides of the prothorax, the scape extends beyond the lateral thoracic spine. In *H. cervinus*, the scape, when in a similar position, does not reach to the thoracic spine; it is scarcely more than half as long as the third joint.

93. **Haplohammus griseipennis**, Thoms. *Monochamus griseipennis*, Thoms. Archiv. Ent. I, p. 296.

Carin Mts. (district of Chebá) alt. 900-1100 m. and Thagatà in Tenasserim. Occurs also in North India.

This species, though of somewhat smaller size, has the general aspect of *H. cervinus*, Hope, and its allies; but it has a distinctly paler pubescence which varies in colour from an almost ashy-grey to a light yellowish-grey; the pronotum and the elytra are more minutely and much more closely punctured; and joints 3rd-5th of the male antennae are not sensibly thickened. The front of the head is slightly transverse, and is minutely and somewhat sparsely punctured; the vertex has also a few minute punctures. The scape of the antennae in the male is a little more than half as long as the third joint. The elytra are rounded at the apex.

94. **Haplohammus basicornis**, sp. n. *Pube griseo-brunnea dense obtectus; capite prothoraceque impunctatis, hoc antice obsolete sulcato, et ad basin recte bisulcato; elytris versus basin sat sparsim punctulato, apicibus rotundatis; antennis (♂) quam corpore fere triplo longioribus, scapo brevi usque ad apicem valde incrassato, articulis $3°$ ad 5um haud incrassatis. Long. 20 mm.*

Carin Mts. (district of Ghecù) at an altitude of from 1300-1400 m.

This species somewhat resembles *H. cervinus*, Hope, but may distinguished by the form of the antennal scape, which is strongly thickened to the apex, and by the absence of any evident thickening of the third, fourth and fifth joints. The joints of the antennae from the third are testaceous in colour with a slight infuscation at the tips. The pubescence, which in the present condition of the type specimen is dull, would probably in the fresh state exhibit a silky or satiny lustre.

95. **Haplohammus admixtus**, sp. n. *Parvus, pube griseo-brunnea sat dense obtectus; capitis fronte vix punctata, sat angusta, quam latitudine distincte altiori,* tuberculis antenniferis anguste separatis; *prothorace supra sparsim punctato; scutello* cinereo: *elytris sat dense punctatis, griseo-brunneo-pubescentibus, versus* medium spar*sim irregulariterque fusco-maculatis; antennis* (♂) quam corpore *triplo longioribus, scapo brevi obconico, apice late* cicatricoso, arti*culo tertio quam scapo plus duplo longiori,* haud incrassato. Long. 11-14 *mm.*

Carin Mts. (district of Chebä), alt. 900-1100 m.

This species is to be distinguished from the preceding by the less uniform pubescence of its elytra, the general greyish-brown tint of which is diversified by some irregular spots and points of a somewhat dark-brown colour; and, more particularly, by the narrowness of the front of its head. The scape of the antennae is rather short and obconical, with a broad and distinct but incompletely margined cicatrix at the apex; the third and following joints are slender, the third is more than twice as long as the scape, the fourth is manifestly, the fifth scarcely, shorter than the third.

96. **Macrochenus Guerini**, White. — *Pelargoderus Guerini*, White. Ann. and Mag. N. Hist., Ser. 3, Vol. II (1858) p. 274.

Bhamò; Catcin Mts.; Carin Mts. (district of Chebä), alt. 900-1100 m.

The species occurs also in Assam, in Siam and in China. The genus, to which it belongs, was first characterized by

Guérin-Méneville (Voy. Delessert, Vol. II (1843), p. 59), who proposed for it the name *Macrochenus*, Pascoe subsequently (Proc. Zool. Soc. 1866, p. 252) described it under the name of *Mecotagus*, by which it has since been generally known. *M. tessellatus*, Guér. (l. c.), the type of the genus, is probably identical with, or, at most, a variety of *M. tigrinus*, Oliv.

97. **Melanauster zonator**, Thoms. — *Cyriocrates zonator*, Thoms. Rev. et Mag. de Zoologie, 1878, p. 50. — *Melanauster zonator*, Ritsema, Notes from the Leyden Museum, Vol. XII, p. 180.

One example taken at Carin Mts. (district of Ghecù) at an altitude between 1300 and 1400 metres. The species occurs also in Siam.

The pale green scaly pubescence of the elytra in this example, instead of being arranged in regular transverse bands, as in the type (with which, through M. René Oberthür's kindness, I have been enabled to compare it) is broken up into spots, of which the first series consists of three, oblong or quadrate, near the base of each elytron, the second of two on each elytron — one, smaller, lateral, the other larger and transverse, near the suture; the third row is made up of one large transverse spot towards the side of each elytron, and of two smaller spots on the left elytron and one on the right elytron, near the suture; the fourth row consists of two spots on each elytron — one, large, transverse, towards the sides, one, smaller, close to the suture. This row is succeeded by a large spot at the apex of each elytron. In its remaining characters the specimen agrees well with the type.

There is an example from Siam in the British Museum collection, in which the third and fourth bands of the right elytron are broken into spots near the suture, and exhibit a want of symmetry in relation to the bands of the left elytron. It is probable, therefore, that a considerable amount of individual variation exists with regard to the arrangement of the pubescence in this species.

98. **Melanauster Medenbachii**, Ritsema. Notes from the Leyden Museum, Vol. III, p. 39 and Vol. XII, p. 180.

I have no hesitation in referring to this species two fine male examples which were taken at Thagatà in Tenasserim.

The pubescent bands of the elytra in these examples are of a beautiful golden green colour. Mr. Ritsema's supposition that the grey colour of the pubescence in his specimen was « due to their having been preserved for some time in spirits » is thus shown to be correct. The characters by which the species has been stated to differ from *M. zonator* are clearly evidenced in the two examples before me.

99. **Melanauster chinensis**, Forst. — *Cerambyx chinensis*, Forst. Nov. sp. Insect. (1771), p. 39.

Catcin Cauri Mts. in Upper Burma.

100. **Aristobia birmanica**, sp. n. A. reticulatori *similis et affinis, sed differt articulis antennarum quarto quintoque apice haud villosis*.

Taken on the Catcin Mts.; at Carin Mts. (district of Asciuii Chebà) at an altitude of 1200-1300 m., at Rangoon in Lower Burma, and at Bhamo in Upper Burma.

This species is allied to, and closely resembles, *A. reticulator*, Fabr. It has, however, somewhat smaller ochreous spots on the elytra and differs, especially, by the complete absence of tufts of black hairs from the fourth and fifth joints of the antennae. The third joint only is furnished with a thick tuft of black hairs. The species resembles *A. clathrator*, Thoms. in the smaller size of the ochreous spots of the elytra.

I have seen a considerable number of examples of this species from Burma, and in all, without exception, the fourth and fifth joints of the antennae have the characters mentioned above.

101. **Aristobia Voeti**, Thoms. Rev. et Mag. de Zool. 1878, p. 51. — *Aristobia angustifrons*, Gahan. Ann. & Mag. N. Hist., Ser. 6, Vol. 1, p. 276.

Carin Mts. (district of Chebà), alt. 900-1100 m. Occurs also in Siam and in China.

Since I described this species under the name quoted above I have seen examples in Bates's collection thicketed *Aristobia Voeti*, Thoms., and M. René Oberthür, during a visit to London, recognized Thomson's species in my type. All the examples I have seen have a pale grey or ashy-grey pubescence, not a yellowish pubescence, as in the specimens described by Thomson. In his description also the prothorax is said to have some yellow spots above, but I think I have more correctly described it as grey (or yellowish) with numerous small black spots above. The difference in coloration is doubtless to be attributed, as M. Oberthür suggests, to the action of the spirit in which some of the specimens were preserved.

102. **Aristobia horridula**, Hope. — *Lamia horridula*, Hope, in Gray's Zool. Miscellany (1831), p. 27. — *Cerosterna fasciculata*, Redtenb. Hüg. Kaschm. IV, 2, (1848), p. 552, pl 27, fig. 2. — *Aristobia fasciculata*, (Red.) Lacord. Gen. des Coléoptères, Atlas, pl. 97, fig. 5.

Carin Mts. (district of Chebà), alt. 900-1100 m.

103. **Coelosterna plagiata**, White. Proc. Zool. Soc. 1858, p. 403.

Palon in Pegu. One example.

104. **Coelosterna carissima**, Pasc. — *Monohammus carissimus*, Pasc. Trans. Ent. Soc. Ser. 2, Vol. IV (1857), p. 104. — *Coelosterna tessellata*, White. Proc. Zool. Soc., 1858, p. 404. — ? *Coelosterna maculosa*, Thoms. Syst. Ceramb. App. p. 552. — *Cerosterna ocellata*, Nonfr., Berl. Ent. Zeit. XXXVI, p. 379.

Meetan in Tenasserim. One example.

105. **Uraecha chebana**, sp. n. U. angustae *affinis; fusco-brunnea; griseo-pubescens; elytris utrinque plaga elongata sordide alba, maculam magnam brunneo-velutinam includente; prothorace lateraliter breviter spinoso, supra sat denseque nigro-granuloso, griseo-pubescente, maculis parvis* quinque *fulvis; elytris basi et medio prope suturam dense sat fortiterque punctatis; ad humeros sub-asperatis.* Long. 18-20 mm.

Taken at the Carin Mts. (district of Chebà), at an altitude of 900-1100 m.

Head and prothorax with a rather thin greyish pubescence. Prothorax with numerous rather closely placed black granules which extend over almost the whole upper surface except the anterior and posterior borders. Elytra closely and rather strongly punctured at the base and along the middle, the punctures disappearing posteriorly: each with a somewhat roughly triangular velvety brown patch at the side just behind the middle; this patch bounded in front and behind by an oblique dirty white patch which appears to be almost impunctate. The whole forepart of the elytra extending triangularly backwards between the lateral plagae is dark brown in colour, and is clothed with a faint greyish pubescence, with, here and there, a sprinkling of fulvous. The posterior part of the elytra has a greyish pubescence irregularly speckled with brown. The body underneath and the legs have a greyish pubescence, with minute rounded naked points wich appear reddish brown or dark brown according to the colour of the derm. The mesosternal process is distinctly, though not strongly, tubercled in the middle.

This species is somewhat closely allied to *U. angusta*, Pascoe, but in the latter the velvety-brown lateral patch of the elytra is slightly darker in colour, is obliquely placed, and has a distinctly, though sparsely, punctured patch clothed with pale greyish or cinereous pubescence, both in front of and behind it. The anterior part of the elytra is as thickly covered with pubescence as the rest of the surface, so that the derm of that part is almost completely concealed. The mesosternum is very faintly or not at all, tubercled.

106. **Uraecha thoracica**, sp. n. *Elongata, modice angustata, pube albo-cinerea, brunnea et fulva vestita; prothoracis disco area subcordata paullo elevato, cujus apice scutellum fere attingente, antice leviter bituberculata, et medio utrinque nigro-granulosa, prothoracis lateribus sat valde spinosis, et ante spinam tuberculo parvo munitis; scutello albo-cinereo; elytris versus basin subasperato sat fortiterque punctatis, punctis versus apicem evanescentibus; apicibus sub-truncatis, fere rotundatis, corpore subtus pedibusque brunnescentibus vel obscure grisescentibus, maculis ep sternorum abdominisque et annulis angustis femorum albescentibus, exceptis. Long.* 14-20 mm.

Taken at Carin Mts. (districts of Chebá and Ghecú), altitude 900-1400 m. and also at Plapoo in Tenasserim.

Head dull greyish or brownish in front, pale fulvous around the eyes and on the occiput, with a median impressed black line extending from the epistome to the thorax, with two short oblique black lines on the vertex, each commencing at, and appearing as a continuation of, the upper lobe of the eye. Prothorax furnished at the middle of each side with a tolerably strong spine, and just in front of the spine with a small tubercle; the disk exhibiting a slightly raised area, somewhat cordate in shape, of which the obtuse apex almost touches the scutellum, and at the anterior extremity of which there is on each side a very small tubercle; along each side of the middle line there are some closely placed black granules (which in the largest specimen appear as two distinct black patches), and there are a few scattered granules extending on to the lateral part of the disk and around the base of the lateral spine and tubercle; the pubescence of the prothorax is partly pale fulvous (chiefly on the raised central area) but mostly dull greyish, with a very small blackish spot surrounded by a whitish patch placed on the side anterior to, and just above, the angle of the cotyloid cavity. The scutellum is whitish, with a very small black point in front. The elytra are somewhat asperately, but not densely punctured at the base; the punctures, becoming gradually feebler posteriorly, entirely disappear at about the beginning of the apical fourth. The velvety brown pubescence of the elytra is limited to a larger irregular and somewhat broken patch occupying the middle of the basal fourth, and to three, or more, patches on each side, of which the one the nearest the middle is much larger than the rest. The body underneath and the legs vary from greyish-brown to brown, with a distinct spot on each of the meso- and meta-thoracic epimera, some small obscure spots on the abdomen, and a narrow sinuous ring near the apex of each of the femora, whitish. The antennae, more than twice as long as the body in the male, and half as long again as the body in the female, are brownish, with a fulvous spot near the outer extremity of the scape, and a dull greyish ring at the base of the third and following joints. The mesosternum is very feebly tubercled between the middle coxae.

107. **Blepephaeus succinctor**, Chevr. — *Monohammus succinctor*, Chevr. Rev. et Mag. de Zool., 1852, p. 417. — *Monohammus sublineatus*, White. — *Monohammus obfuscatus*, White.

Taken at Thagatà in Tenasserim ; also occurs in China, Assam and Malaya.

108. **Blepephaeus stigmosus**, sp. n. *Pube brunnea vel cervina sat dense vestitus; capite impunctato, oculis magnis; prothorace supra sparsim punctato, disco utrinque obsolete bituberculato; elytris basi et plaga utrinque pone medium infuscatis, haud dense punctatis, punctisque majoribus dispersis fusco limbatis ornatis; apicibus truncatis; antennis brunneo-testaceis, quam corpore (σ) plus sesqui-longioribus. Long.* 18-20. *Lat.* 5 $1/_2$-6 *mm.*

Tenasserim, Thagatà and Meetan.

Head impunctate, with a brownish pubescence; the lower lobes of the eyes large and somewhat rounded, reaching below almost to the base of the mandibles. Scape of the antennae rather long and cylindrical, equal in length to two thirds of the third joint; eleventh joint a little longer than the tenth. Prothorax with a thick brownish pubescence, sparsely punctured above and with two very feeble tubercles on each side above; scutellum cervinous. Elytra with a light brown or fawn-coloured pubescence, with the base and a somewhat triangular patch on each side behind the middle infuscate, with minute rounded fuscous spots which are sparsely scattered, and in the centre of each of which there is a rather large puncture. The apices of the elytra are truncate. The middle tibiae are without groove or tubercle. The mesosternum is feebly tubercled between the middle coxae.

This species, which is smaller and narrower than *Blepephaeus succinctor*, Chevr. appears to be nearly allied to *Blepephaeus subcruciatus*, White (*Monohammus*) and still more closely to an undescribed form from North India.

109. **Eutaenia Oberthuri**, sp. n. *Cinereo-pubescens; prothoracis fascia transversa media, elytrorumque fasciis tribus communibus et macula rotunda utrinque ante apicem nigris; antennis ($\female$) quam corpore paullo longioribus, nigris, articulis 3°-7° basi cinereis. Long.* 21-23 *mm.*

Carin Mts. (district of Chebá), alt. 900-1100 metres.

Cinereous; with a transverse black band on the prothorax which extends over, and embraces, the tubercle on each side, with three somewhat irregularly formed transverse black bands on the elytra and a rounded spot of the same colour on each a little before the apex. The female antennae surpass the apex of the elytra by about the last three or four joints; joints 3-5 are subequal in length, and each is longer than the sixth or seventh; joints 8-10 are also subequal, and each much shorter than the seventh; the eleventh is scarcely longer than the tenth.

The last abdominal ventral segment of the female presents a most unusual character. This segment is rather broadly emarginated at the apex, the emargination being limited on each side by a short blunt tooth; at a short distance in front of the apical margin there projects from the ventral surface a lamellar process, bidentate at its free extremity, with a narrow angular emargination between the teeth; a narrow transverse depression somewhat wider in the middle, is enclosed between this process and the downwardly curved apical portion of the segment, and is thickly lined with fulvous hairs, the arrangement being such that the depression itself might readily be mistaken for the terminal opening. An exactly similar structure exists in the females of *Eutaenia trifasciella*, White — the type of the genus. The latter species differs from the one above described by having distinctly shorter antennae (in the female at least), by the tawny-yellow colour of its pubescence, and its narrower black bands.

Eutaeniopsis, Gahan, is a synonym of *Eutaenia*, Thoms. Thomson's characterization of the genus is brief and somewhat inaccurate. I have seen his type, through the kindness of M. René Oberthür, to whom I am happy to dedicate the new species.

110. **Cycos subgemmatus**, Thoms. — *Monochamus subgemmatus*, Thoms., Archiv. Ent., I, p. 294. — *Monochamus georgius*, White, Proc. Zool. Soc., 1858, p. 407.

Carin Mts. (district of Chebà). One example. The British Museum collection contains examples from Sylhet in Assam, from Siam, Cambodia, and Burma.

111. **Pharsalia antennata**, sp. n. *Fusca, prothorace supra lineis quatuor fulvis; elytris maculis paucis atro-velutinis et maculis minutis fulvis sub-confluentibus et irregulariter dispersis; capite tuberculis antenniferis contiguis; elytris valde punctatis obtuse carinatis, utrisque ad basin tuberculo magno munitis; apicibus rotundatis; antennis quam corpore multo longioribus, rufo-ferrugineis, articulorum apicibus fuscis; articulis 3° et 4° ($\male$) apice incrassatis articulis 3° et 4° ($\female$) apice nodosis. Long. 20-25 mm.*

Upper Burma (Catcin Cauri) one $\male$ example, and Carin (district of Chebà) one $\female$ example.

Dark brown, prothorax with four fulvous lines above; elytra with some velvety black spots and with numerous minute fulvous and grey spots which here and there run together. Head with the inner horns of the strong and vertical antennal tubercles meeting in the middle line, and leaving a slight depression between them only on the anterior face. Prothorax slightly rugose above, and with a sharp conical tubercle on each side. Elytra rather coarsely punctured, especially towards the base, strongly nitid where not covered with pubescence, each with two or three obtuse carinae, and furnished with a strong tubercle, whose anterior face rises sub-vertically from the basal margin; apices of the elytra rounded. Body underneath with minute scattered fulvous spots. Legs reddish brown with faint grey pubescence. Antennae in the female more than half as long again as the body (the antennae in the single male example are broken off from the seventh joint), with the third and fourth joints in the female nodose at the apex, with the same joints in the male thickened, but not nearly so strongly as in the female; the scape is ferruginous brown, the remaining joints reddish brown with their apices fuscous.

The exceptional characters of this species will enable it to be readily recognized. Notwithstanding the close contact of its antennal tubercles, and the decided thickening of the apices of the third and fourth joints of the antennae, I do not see that the species can be better placed than in *Pharsalia*.

112. **Batocera rubus**, Linn. — *Cerambyx rubus*, Linn., Syst. Nat. Ed. X, p. 390.
Catcin Cauri, Upper Burma.

113. **Batocera albofasciata**, De Geer. — *Cerambyx albofasciatus*, De Geer., Mém., V. p. 106, pl. 13, fig. 16.
Carin (district of Chebà), alt. 900-1100 m.

114. **Batocera Roylei**, Hope, Trans. Zool. Soc., I, p. 103, pl. 15, fig. 1.
Carin (district of Chebà), alt. 900-1100 m.

115. **Batocera Calanus**, Parry, Trans. Ent. Soc., Vol. IV, p. 86.
Carin (district of Chebà), alt. 900-1100 m.

116. **Batocera adelpha**, Thoms., Archiv. Entom., I, p. 77.
Carin Mts. (district of Chebà).

117. **Batocera Titana**, Thoms., Arcana Naturæ, p. 82.
Carin (district of Asciuii Chebà), alt. 1200-1300 m.

118. **Apriona rugicollis**, Chevr., Rev. Zool., 1852, p. 418.
Carin (district of Chebà). One example.

119. **Calloplophora Sollii**, Hope. — *Oplophora Sollii*, Hope, Trans. Linn. Soc. XVIII (1841), p. 438, pl. 30, fig. 4.
Carin (district of Chebà). One example.
This fine species occurs also in Assam, and at Darjeeling.

120. **Himantocera penicillata**, Hope. — *Lamia penicillata*, Hope, in Gray's Zool. Misc., p. 28.
Taken at Thagatà in Tenasserim, and at Bhamò in Upper Burma.
The species occurs also in North India (Sylhet, Darjeeling, and Nepal).

121. **Himantocera vicina**, sp. n. H. penicillatae *persimilis sed differt articulo antennarum quarto inter medium apicemque non constricto, et articulis* 8° *ad* 10um *apice haud fasciculato-pilosis.*
Carin (districts of Chebà and Ghecù), alt. 900-1500 m.
This species so closely resembles *H. penicillata*, Hope, in size, general form and style of coloration, that only the points of difference need be stated. The fourth joint of the antennae is

gradually thickened from before the middle up to the apex; the joints succeeding the fourth are all in both sexes destitute of small tufts of hairs, and, with the exception of the fifth, have an uniform fulvous-grey pubescence. This last character, however, though holding good for the few specimens from the Carin Mts. which I have seen, is liable, I think, to variation and may be considered of very little importance.

In *H. penicillata*, Hope, the fourth joint of the antennae is thickened at the middle, and, between this thickening and the large swelling at the apex, is somewhat narrowed or constricted; in the males of this species a small but distinct tuft of black hairs is to be seen on the underside at the apex of each of the joints from the eigth to the tenth; a few black hairs of similar character are to be found in fresh specimens at the apex of the seventh and at the middle of the eleventh joint.

In specimens from Java and Borneo which I refer to *H. plumosa*, Oliv., the fourth antennal joint has a form similar to that of *H. penicillata*, but the succeeding joints are in both sexes destitute of small pencils of hairs. In these specimens also the disk of the prothorax is a little less uneven than in *H. penicillata* or *H. vicina*.

So far as the relative length of the prothorax is concerned I am unable to detect any appreciable differences in the three species just mentioned. What Lacordaire regarded as a specific distinction seems to me to be simply a sexual difference. In the males of all the species the prothorax is decidedly longer than in the females.

122. **Golsinda basicornis**, sp. n. G. corallinae *persimilis sed minor, magis parallela, scapo antennarum apice minus abrupte incrassato, quam articulo tertio paullo breviore.* Long. 14-22 mm.

Carin (district of Ghecù), alt. 1300-1400 m.; also Siam, Laos, and Allahabad (British Museum collection).

This species strongly resembles *Golsinda corallina*, Thoms. with which it seems to have been confounded by Mr. Pascoe (vide Longic. Malay. p. 133) as it was by myself until a recent examination had shown me that the specimens from Laos and

Siam are specifically distinct from those from Borneo. The elytra in the new species are more parallelsided, the scape of the antennae is less elongated, and less abruptly thickened at the apex, is rather shorter than the third joint, and carries a reddish pubescent patch near the middle. In *corallina* the elytra are considerably narrowed from the base backwards, the scape of the antennae is rather abruptly thickened at the apex, is slightly longer than the third joint and has not a reddish patch.

123. **Mesosa** (¹) **subfasciata**, sp. n. *Pube fulvo-brunnea griseaque, fusco-maculata, obtecta; elytris fascia transversa media, lata sed irregulariter marginata, pallidiore, ante hanc fasciam fere totis fuscis; apicibus anguste rotundatis; antennis fuscis, scapo griseo, articulis 3° et sequentibus basi cinereis. Long.* 11-15 *mm.*

Carin Mts. (district of Chebá), alt. 900-1100 m.

This species has the head slightly less concave, the prothorax more rounded at the sides, and the elytra more attenuate posteriorly than in *Mesosa nebulosa*, Fabr., but otherwise agrees with it fairly well in structural characters. The antennae of the male are more than half as long again as the body. In this sex also there is a narrow transverse pilose depression towards each side of the anterior border of each of the three intermediate abdominal sternites. Each depression is concealed more or less by a fringe of fulvous hairs attached to the hind margin of the sternite immediately in front. The antennae of the female reach to a little beyond the apex of the elytra.

124. **Mesosa obscura**, sp. n. *Griseo et fulvo-griseo pubescens; punctis maculisque fusco-brunneis; capite inter antennas sat fortiter concavo; prothorace supra leviter trituberculato, basi apiceque recte lineato-sulcato et paullo pone apicem sinuato-impresso, lateribus leviter inaequalibus et obsolete rugosis, elytris punctatis, fulvo-griseopubescentibus, maculis punctisque fusco-brunneis, corpore subtus*

(¹) The genus *Mesosa* was first characterized by Latreille (Règne Animal, 2 de Edit. Tom. V, p. 124); though Serville is usually cited as its author. *Aplocnemia*, Stephens (Illustrations of Brit. Ent. IV (1831) p. 236), which is a synonym of *Mesosa*, was not described until two years later.

pedibusque griseis, femoribus medio tibiisque basi et apice fusco-maculatis; antennis griseis, articulis apice fuscis. Long. 12-15 mm.
Bhamó in Upper Burma. Two examples.

Head strongly enough concave between the antennal tubercles, minutely and rather sparingly punctured in front, feebly impressed along the middle line. Prothorax rather narrow, with three feeble tubercles on the disk, of which the third and smallest is placed near the base (this tubercle is, in the larger of the two specimens before me, divided in the middle by a narrow groove); the sides are made slightly uneven by the presence on each of two very feeble and obtuse tubercles — one a little behind the anterior margin, the other much farther back and a little higher up; between these tubercles and extending upwards on to the disk are some feeble transverse ridges (not evident in the smaller specimen). Scutellum dark brown with a grey centre. Elytra strongly enough and rather closely punctured, the sides nearly parallel; pubescence grey with irregularly scattered small fuscous spots and points. Antennae ($\male$) about one half longer than the body: the first two joints grey, the remaining joints grey at their bases only; first joint not much more than half as long as the third joint; the latter longer than the fourth; the eleventh joint with a short hooked appendix. Intermediate segments of the abdomen with narrow, transverse, pilose depressions placed anteriorly towards each side.

In this species some of the characters of *Coptops* are united with those of *Mesosa*; but they point on the whole to its position in the latter genus.

125. **Cacia ornata**, sp. n. *Griseo-pubescens; capite linea media glabra impresso, vertice lineis duabus ab thorace ad frontem descendentibus et vitta brevi utrinque pone oculum atro-velutinis; prothoracis dorso sparsim punctato vitta mediana fusca et antice maculis duabus atro-velutinis, cinereo-limbatis, ornato; lateribus utrisque macula parva sublineari atro-velutina; elytris sat fortiter punctatis, lineis maculisque atro-fuscis et maculis quinque albis ornatis; antennis ($\male$) quam corpore sesqui-longioribus, griseis,*

articulis 5°, 7° 9°-*que fere totis et apicibus ceterorum* (1° et 2° *exceptis*) *fuscis. Long.* 15 *mm.*

Two examples; one taken at Palon in Pegù, the other at Carin (Chebà).

Pubescence greyish, somewhat paler on the underside of the body. Head with a median impressed glabrous black line; with two velvety black lines which extend from the thorax downwards between the antennary tubercles, slightly diverging in their course, and becoming obliterated on the front of the head; with a short velvety black vitta on each side above behind the upper lobe of the eye. Prothorax somewhat sparsely punctured above, with a median dorsal fuscous vitta, and with two velvety black spots, narrowly margined with cinereous, which are placed one on either side of the anterior part of the disk; with a small longitudinal black spot placed somewhat anteriorly and low down on each side. Scutellum fuscous. Elytra strongly enough, but not very thickly punctured, adorned with numerous lines and about eight small spots of a velvety black colour, and with five white spots. Of the black spots three are placed along each outer margin, and two posteriorly on the suture. Of the white spots, one is placed somewhat dorsally on each elytron a little in front of the middle, another smaller and less distinct behind the middle, while the fifth is on the suture and just in front of the anterior of the two sutural black spots. Legs and underside of body grey, with some fuscous points and spots; abdominal segments fringed with pale fulvous-grey posteriorly. Prosternal process strongly arched, declivous behind; mesosternal process vertical in front, feebly tubercled near the lower anterior border. Antennae (♂) about one half longer than the body, ciliate underneath, the ciliae being more densely aggregated and blackish in colour beneath the fifth and the apex of the third joint.

This rather pretty species may be easily recognized by its peculiar markings which I have described above in some detail.

126. **Cacia cretifera**, Hope. — *Lamia cretifera*, Hope, in Gray's Zool. Misc. (1831), p. 27.

Thagatà in Tenasserim, Teinzò in Upper Burma, and Carin (district of Chebá); occurs also in Nepal, Sikkim, Assam, Andaman Islands and South China.

127. **Agelasta nigromaculata**, sp. n. (Pl. I, fig 9). *Griseo-pubescens, capite prothoraceque nigro-vittatis, elytris nigro maculatis et punctatis; antennis nigris articulis 3° ad 6um basi cinereis; pedibus nigro annulatis et punctatis; prothoracis lateribus antice obsolete tuberculatis; mesosterno antice verticali et leviter tuberculato; prosterno postice verticali; articulo tertio antennarum quam primo paullo longiori*. Long. 16. Lat. 7 mm.

Malewoon in Tenasserim. One example.

With a yellowish-grey pubescence. Head impunctate, feebly concave between the antennal tubercles, with two longitudinal black vittae extending almost from the labrum in front to the occiput behind, and with a short black vitta or spot behind the upper lobe of each eye. Prothorax slightly rounded at the sides, each of which bears a small and almost obsolete tubercle placed a little behind tho front margin; with two black vittae on each side, one broad placed low down, the other, very much narrower and almost lineate, placed higher up towards the disk, the disk also with a black vitta on each side and with three somewhat glabrous spots surrounding the middle, the posterior of which appears very slightly elevated and divided by a feeble median impressed line. A few black punctures are scattered through the fulvous-grey bands which occupy the intervals between the black vittae. The scutellum is black, with a grey spot in the middle, and is surrounded by a somewhat rectangular black spot common to the two elytra; another common oblong spot is placed on the elytra just before the middle, while on each elytron there are about eight other somewhat similar but more irregular spots, of which the largest lies at, and below the shoulder, the elytra bear in addition some scattered black points in the centre of each of which lies a puncture. The body underneath has a yellowish-grey pubescence interrupted by black spots and points. Legs grey, with black small rounded points on the femora, a ring at the apex of each femur and two on each tibia black. Antennae (♂) about half as long again as the body, shortly and thickly ciliate below, black, with the scape slightly, the bases of the joints from the third to the sixth, and the tip of the last joint pale grey; third joint slightly longer than the first or fourth.

128. **Agelasta mixta**, sp. n. *Brunnescente-pubescens, capite prothoraceque nigro-vittatis; elytris nigro cinereoque maculatis et nigro-punctatis, punctis praecipue versus basin in seriebus longitudinalibus aggregatis, corpore subtus pedibusque fulvo-brunneo-pubescentibus, femoribus sparsim nigro-punctatis, tibiis apice nigris; antennis griseo-brunnescentibus, articulis 3° ad 6um apice, et 7° ad 10um totis nigro-fuscis.* Long. 17-18 mm.

Carin Mts. (Chebà district).

Prothorax with two feeble nearly obsolete tubercles on each side, one at the middle, the other a little behind the anterior margin. Third joint of the antennae distinctly longer than the first or fourth. Prosternal process strongly enough arched and rounded in the middle, vertical behind. Mesosternal process vertical in front, horizontal below, feebly tubercled at its antero-inferior edge.

This species seems nearly enough allied to the preceding, but may be easily distinguished by the general brownish tone of colour of its pubescence, the fewer black spots on the elytra, and the presence of three ashy spots along the suture and some scattered spots on each elytron. The black bands of the head and prothorax are disposed in very much the same manner as in *C. nigromaculata*, the black points on the elytra are more numerous, and arranged, especially towards the base, in somewhat irregular longitudinal series.

The two species just described agree with *A. transversa*, Newm. — the type of the genus — in the relative proportions of the joints of the antennae, in the shape and tuberculation of the prothorax, and of the pro-and mesosternal processes; they agree, in fact, in all essential points of structure. Their relatively narrower elytra, however, give them an appearance more resembling the species of *Coptops;* but from the more typical species of the latter genus they are be distinguished by the absence of a distinct antero-lateral tooth or tubercle from the prothorax, by having the posterior face of the prosternal process vertical, and the third joint of the antennae longer than the first or fourth. In the males each of the three intermediate ventral segments of the abdomen has on each side anteriorly a narrow transverse depression which is more or less concealed by the fringe of hairs attached to the hind margin of the immediately preceding segment. This character is not present in the males of the typical species of *Coptops*, nor, with the exception of *A. transversa*, Newm., *A. bifasciana*, White, and possibly a few others, does it exist in those species hitherto included in *Agelasta*.

129. **Coptops annulipes**, sp. n. *C. fuscae sat similis et affinis, sed differt pube rufo-mixta; prothoracis dorso distinctius fusco-bivittato, lateribus utrisque vitta nigro-fusca, distincta, paullo supra coxam posita; elytris utrinque medio plaga obliqua, pallide grisescente plus minusve distincta; pedibus nigro-fusco distincte annulatis. Long.* 14-20 mm.

Carin Mts. (district of Chebá), alt. 900-1100 m. The British Museum collection contains examples from Siam, Cambodia, Burma, and North India.

This species is closely allied to *Coptops fusca*, Oliv., of which it may possibly be only a well marked variety. The pubescence has usually a considerable admixture of red. The head has two dark-brown vittae above; there are two dark-brown vittae also on the disk of the prothorax, while on each side, just above the coxa, there is another very distinct blackish brown vitta which does not quite reach to the anterior or posterior margin. The elytra are marked with spots and points of dark-brown; while a pale ashy-grey pubescence, slightly mixed with red, forms a more or less distinct and broad, oblique fascia across each elytron near the middle, and extends forwards for some distance in the sutural region. The same characters that serve to distinguish this species from *C. fusca* will distinguish it also from *Coptops aedificator*, Fabr. The two latter in fact are scarcely distinct.

Coptops aedificator, Fabr. is sometimes known under the name of *bidens*, Fabr., with which it is placed as a synonym in the Catalogue of Gemminger & Harold. But the characters of *Lamia bidens*, Fabr. — « thorace acute-spinoso. Elytris apice bidentatis » — show that this species does not even belong to the genus *Coptops*. The mistake probably arose from the fact that a specimen of *C. aedificator* in the Banksian Cabinet is erroneously ticketed *Lamia bidens*, Fabr. To this specimen there is an additional label attached, bearing the words « an bidens Fabricii? ». The actual type of the species appears to have been lost.

130. **Coptops Pascoei**, sp. n. *Brunneo-griseo-pubescens, punctis maculisque fuscis; capitis fronte infuscata; prothorace supra medio leviter tri-tuberculato, lateraliter utrinque vitta lata fusco-velutina; scutello medio griseo, lateribus fuscis; elytris punctis maculisque fuscis, his fascias irregulares duas, vel tres formantibus; corpore subtus griseo sparsim fusco-punctato; pedibus griseis, fusco-annulatis; antennis articulis 2° et 3° fere totis, ceterisque apice infuscatis. Long. 11-17 mm.*

Bhamô in Upper Burma (*Fea*); Siam, Cambodia, (in coll. Brit. Museum).

Pubescence greyish, varying, especially on parts of the elytra, to a pale brown. Head somewhat dark-brown in front. Prothorax grey to greyish-brown, with a broad, dark-brown, velvety band on each side extending from the anterior to the posterior margin, and with its upper and lower margins somewhat irregular, the disk with three feeble tubercles, the sides each with a small tubercle placed a little behind the anterior margin and at the upper border of the fuscous-band, with a second, somewhat larger, but more obtuse, tubercle a little higher up and nearer the middle line. Scutellum grey, bordered at the sides with fuscous. Elytra punctured, each with a very small tubercle near the middle of the basal margin, and with an obtuse prominence a little farther back, pubescence grey and pale brown with scattered points, and irregular spots, the latter aggregated to form indefinite dark-brown bands, of which one is just behind the basal prominences, and is followed by an isolated spot on each elytron, the second is a little behind the middle, while a third very indistinct band is placed a little before the apex.

This species, which is allied to, and somewhat closely resembles *Coptops aedificator*, Fabr., *C. fusca*, Fabr., and other species, may yet be sufficiently distinguished by the broad and distinct dark-brown band on each side of the prothorax, taken together with the other characters detailed above. I have named it in memory of Mr. Pascoe to whose labours we are indebted for the descriptions of most of the known species of this difficult group.

The species has hitherto been known only from Siam and Cambodia.

131. **Coptops vomicosa**, Pasc., Journ. of Ent., I, 1862, p. 341.
Carin Mts., Chebà district, 900-1000 m.

Mesolophus, g. n.

Head strongly enough concave between the antennal tubercles. Eyes sub-divided. Antennae (♀) longer than the body; the scape elongate, a little shorter than the third joint; the latter slightly curved, longer than the fourth, the following joints gradually decreasing in length. Prothorax transverse, unarmed at the sides. Elytra moderately long, subacute and minutely tubercled at the shoulders, from thence narrowed posteriorly; apices broadly and somewhat obliquely truncate, with the outer angles obtusely rounded off; each elytron with a short prominent crest placed a short distance back from the middle of the base. Prosternal process with its posterior face vertical, its postero-inferior edge, which is somewhat obtusely pointed in the middle, projecting slightly backwards; the mesosternal process is strongly declivous or almost vertical in front, its ventral face is very short and is slightly raised along the middle. The legs of the female are sub-equal in length; the intermediate tibiae entire; the claws of the tarsi divergent.

This new genus of the *Mesosides* resembles *Aesopida* in having a basal crest on each of the elytra, but this crest does not reach as far as the anterior margin; it is distinguished further from *Aesopida* by the rather acute shoulders of its elytra, and by the absence of tubercles or teeth from the sides of the prothorax.

132. **Mesolophus humeralis**, sp. n. (Pl. 1, fig. 10). *Niger, pube fulvo-brunnea, fasciis maculisque atro-velutinis ornatus, antennis nigris, articulis 1° 2°que brunnescentibus, ceteris basi cinereis. Long. 16 mm.*

Carin Mts. (Chebá district), alt. 900-1100 m. One example.

With a fulvous brown pubescence, here and there interrupted by cinereous and fuscous spots. Head with two oblong fuscous spots above. Prothorax slightly uneven on the posterior part of the disk, with two somewhat interrupted velvety black vittae, one towards each side of the disk, above. Elytra each with two zig-zag velvety black bands, of which the first, somewhat broader, extends from the basal crest to reach the outer margin a little farther back, the second, a little behind the middle, passes from the outer margin to join the suture, where it forms, with its fellow of the opposite side, an oblong spot lying behind the fascia. Between the second band and the apex there are on each elytron two distinct fuscous or black spots, placed obliquely, the outer being slightly posterior to the inner. In addition to these more definite bands and spots there are some smaller fuscous spots and points, and a few cinereous spots on the elytra. The body underneath and legs have a fulvous brown pubescence; the former is somewhat spotted, and the latter ringed, or spotted, with fuscous.

133. **Aesopida malasiaca**, Thoms., Syst. Ceramb., p. 62; Lacord. Gen. Atlas, pl. 99, fig. 4.

Bhamò in Upper Burma.

134. **Palimna annulata**, Oliv. — *Cerambyx annulatus*, Oliv. Entomologie, IV, n. 67, p. 96, pl. 20, fig. 151.

Toungoo, Teinzò and Bhamò in Upper Burma.

135. **Thysia Wallichii**, Hope.

Teinzò, Bhamò in Upper Burma; Carin Mts. (Chebá district) 900-1000 m.; Thagatà.

136. Xylorrhiza adusta, Wied. — *Lamia adusta*, Wied. Zool. Mag., I, 3, (1819) p. 182. — *Xylorhiza venosa*, Casteln. Hist. Nat. des Insectes, II, p. 476.

Upper Burma (Bhamò) and Carin (district of Chebá), altitude 600-800 m.

The males of this species possess a secondary sexual character which I have not seen noticed in any description of the species. This consists of a large transverse oval depression on each side of the third, fourth and fifth abdominal segments. Each of the depressions is lined with rather long hairs.

137. Ioesse sanguinolenta, Thoms. Syst. Ceramb., p. 68; Lacord. Genera des Coleopt., IX, p. 449.

Two examples (σ and φ) were taken by Mr. Fea at Carin Mts. (Ghecù district); alt. 1300-1400 m.

Lacordaire in his characterization of this genus has given a very erroneous description of the antennae.

With the exception of the first joint, which has a cinnabar-red pubescence, like the head, thorax, &c., the antennae are closely covered with a short dull-black pubescence, and the joints from the third to the sixth or seventh have a short fringe of black hairs underneath. In the male (hitherto undescribed) the antennae reach to about the apex of the elytra, in the female they scarcely surpass the middle of the elytra. In neither sex do they bear the elongate fossae which Lacordaire specially mentions; nor have they any punctures beyond the extremely minute pits from which the hairs of the pubescence spring, and which only become visible when the latter is rubbed off. The male differs also from the female in having a rather wide angulate emargination at the apex of the last abdominal ventral segment. As Lacordaire's description of the female agreed so well in every other respect with the female specimen before me, I was unable to account for the remarkable discrepancy in the characters of the antennae. Mr. René Oberthür, by very kindly sending me the specimen which he believed to be the original of Lacordaire's description, has enabled me to explain the anomaly. In this specimen three joints only of the antennae — the two first on one side, and one on the other — belong to the insect; the remaining and chief parts of the antennae were very neatly fastened on to these basal joints, and were probably originally taken from a female specimen of *Morimus funereus*, Muls. The antennae thus manipulated, looked extremely like the true antennae, and might readily deceive one who was not examining them especially in order to detect something wrong. It is curious, nevertheless, that Lacordaire, though calling special attention to the character of the antennae in this genus, has altogether omitted to notice the same character when treating of the genus *Morimus*.

138. **Rhodopis aberrans**, sp. n. *Nigra, pube fulvo-grisca obtecta; prothorace antice posticeque sulcato-constricto, lateraliter medio unidentato, dorso sparsim punctato, antice leviter binodoso, utrinque vittis duabus fuscis; elytris dense punctatis, pube fulvo-grisea fusco plagiata vel marmorata obtectis; antennis (♀) quam corpore fere duplo longioribus, articulo 3° ad apicem sensim incrassato; scapo ad apicem cicatricoso, cicatrice haud distincte marginata. Long.* 20. *Lat.* 6.5 *mm.*

Metanja on the Catcin-Hills, Upper Burma. One ♀ example.

The unique example of this species resembles a very large specimen of *R. pubera*, Thoms. The colour and disposition of the pubescence are almost exactly the same; the chief difference in this respect being in the relative width and distinctness of the fulvous-grey and fuscous vittae of the prothorax; in *pubera* the pale vittae are much narrower than the dark ones, and are well defined; in the present species the dark bands are narrower, and all are ill defined; the median fulvous-grey vitta is marked in the middle by a small glabrous black spot. The disk of the prothorax is much less thickly punctured than in *pubera* and bears a very feeble tubercle on each side just behind the anterior transverse groove. The lateral tooth at the middle of each side is more distinct than in *pubera*. The apex of the antennal scape has a somewhat rounded cicatrice; but this cicatrice is not, as in the Monohammides, limited by a distinct carina. Notwithstanding the presence of the last character, which is quite exceptional in the group, the present species can scarcely, I think, be considered misplaced in the genus *Rhodopis*.

139. **Olenecamptus bilobus**, Fabr. — *Saperda biloba*, Fabr., Syst. Eleuth., II, p. 324. — *Olenecamptus serratus*, Chevr., Mag. de Zool., 1835, p. 134.

Upper Burma (Bhamó), and Tenasserim (Thagatá).

140. **Gerania Boscii**, Fabr. — *Saperda Boscii*, Fabr., Syst. Eleuth., II, p. 323. — *Gerania Boscii*, Serv., Ann. Soc. Ent. Fr. IV, p. 71; Pascoe Longic. Malay., p. 321, pl. XIV, fig. 7 (σ)

Shwegoo in Upper Burma. One example.

141. **Nyctimene agriloides**, Thoms., Archiv. Entom., I, p. 314. One example taken at Thagatá.

142. **Moechotypa thoracica**, White. — *Niphona thoracica*, White, Ann. and Mag. Nat. Hist., Ser. 3, Vol. II, p. 266.

Carin (district of Chebá) and Upper Burma (Catcin Cauri).

143. **Moechotypa umbrosa**, Lacord.? Genera des Coléoptères, Vol. IX, pl. 2, p. 519 (¹).

Two examples taken at Carin Mts. (Chebá district) at an alt. of 900-1100 m.

144. **Moechotypa delicatula**, White. — *Niphona delicatula*, White, Ann. and Mag. N. Hist., Ser. 3, Vol. II (1858) p. 268.

Two examples, one of which was taken at Catcin Cauri in Upper Burma, the other at Carin Mts. (district of Ghecù). The British Museum collection contains examples from Sylhet in Assam and from Laos.

This species is very nearly allied to *Moechotypa suffusa*, Pasc. which it closely resembles in coloration and markings.

145. **Moechotypa verrucicollis**, sp. n. *Pube fusca, ferruginea et grisea vestita, oculis magnis, capitis fronte angustata, prothoracis dorso utrinque tuberculo magno obtuso, lateribus utrisque pone medium tuberculatis; elytris basi granulosis, pilis nigris fasciculatis, utrisque tuberculo parvo in depressione humerale posito. Long.* 22-23 mm.

Upper Burma, Bhamó.

This species may be distinguished from *M. thoracica*, White by its larger eyes, and the consequent narrower front of its head, the more obtuse and somewhat flattened tubercle on each side of the disk of the prothorax; by the small unfasciculated tubercle placed in the supra-humeral depression at the base of each elytron; and by its generally darker coloration.

146. **Habryna Petri**, De Paiva, Ann. and Mag. Nat. Hist., Ser. 3, Vol. VI (1860), p. 360.

Carin Mts. (district of Chebà), alt. 900-1100 m.

147. **Niphona Ferdinandi**, De Paiva, Ann. and Mag. N. Hist., Ser. 3, Vol. VI, p. 361. — *Aelara Ferdinandi* (Paiva) Thoms., Syst. Ceramb., p. 55.

Carin Mts. (district of Chebà). One example.

The example taken belongs to a variety characterized by having a rather large whitish plaga, placed before the middle on the side of each elytron. In the examples of this species in the British Museum collection which have been brought from Cambodia, there is in a similar position, an indistinct plaga formed of coalesced spots or small patches of a rust colour like that of the general pubescence.

The males of this species are provided with a distinct tooth on the ventral side of the anterior tibiae a little below the middle of their length, and also with a small tubercle or tooth at about the middle of the dorsal border of the anterior femora. The females sometimes exhibit a trace of a tooth on the anterior tibiae in a similar position to that of the male. The last ventral segment of the abdomen has almost exactly the same form in both sexes, but in the female is impressed along the middle, with a feeble groove. There is a more distinct fringe of fulvous hairs, attached to the hind margin of the first abdominal segment in the male: towards the sides this fringe conceals a very narrow transverse depression or declivity at the anterior margin of the succeeding segment; but there is on the segment no distinct semi-oval and piligerous depression as in *N. picticornis*, Muls. and certain other species.

The anterior tarsi of the males do not seem to be appreciably more dilated than those of the females.

148. **Niphona vicina**, sp. n. N. furcatae (Bates) persimilis et affinis, sed elytrorum apicibus minus obliquiter truncatis. Long. 15-16 mm.

Thagatà in Tenasserim. Two examples.

Clothed with a pubescence which is mostly of a yellowish or ochreous-grey colour, but which is ashy-white on the inner side of the legs, along the middle of the underside of the body, and on the sides of the elytra. Head with the lower lobes of the eyes moderately large, the vertex forming with the front an almost continuous curve. Prothorax very slightly rounded and unarmed at the sides; the disk longitudinally ridged in the middle, with, usually, three of the ridges regular and distinct, while those external to them are broken and irregular, so that the outer part of the disk and the sides of the prothorax have a somewhat roughly granular appearance; the anterior part (almost a third) of the disk is smooth; while at the base there is a fine transverse groove, followed in the middle by a slight posterior lobe which fits closely against the scutellum. The elytra, broadest at the base, gradually narrow up to the beginning of the apical fourth, and thence are more strongly attenuated to the apex; the apical truncation of each elytron is somewhat oblique with a slight emargination close to the suture; the outline of this margin is more or less obscured by a fringe of hairs; the punctuation of the elytra is strong, but not very dense, and is more evident on a dorsal sub-glabrous area (which may have been produced by rubbing). Each elytron carries at the base a short longitudinal crest, surmounted by yellowish-brown hairs, and a similar, but much feebler crest, in the depression between this and the shoulder.

This species very closely resembles *Aelara furcata*, Bates, and is chiefly to be distinguished by the difference in form of the apices of the elytra. In *furcata* the apices are much more produced on the outside so that the truncation has a strongly oblique direction; in *vicina* the elytra are only slightly prolonged at the outer apex, and the truncation of each is consequently much less oblique.

149. **Niphona Batesi**, sp. n. *Pube ochracea dense vestita, elytris plagis pallidioribus; capite supra inter oculos leviter depresso, lobis oculorum inferioribus magnis; prothoracis dorso sparsim punctato, lateribus sparsim asperato-punctatis, punctis piligeris; elytris ad basin latis, deinde versus apicem valde angustatis, apicibus divaricatis et sub-acuminatis. Long.* 23-28 *mm.*

Carin (district of Chebà). Two examples.

Thickly clothed with an ochreous pubescence; the elytra with four plagae of a paler colour (varying from pale-ochreous to ashy-white) of which one, common, and somewhat triangular is placed at the base, one also common, and having a somewhat irregular hastate form, is placed at some distance behind the first; the remaining two are placed one on the side of each elytron, extending from beneath the shoulder almost up to the beginning of the posterior third, and slightly expanding inwards just behind the shoulder. Prothorax with the anterior margin directly transverse, the posterior slightly lobed in the middle and sinuate towards each side; disk sparsely punctate across the middle; sides sparsely punctured with each of the piligerous punctures placed at the summit of a small granule. Elytra rather sparsely punctured, and with a few larger scattered punctures each occupying the centre of a minute fuscous spot. The apices of the elytra are cut away very obliquely from the suture, and being attenuated also on the outside, each ends in an obtuse point which appears to diverge from its fellow.

150. **Niphona princeps**, sp. n. (Pl. I, fig. 11). *Precedenti similis, sed elytris brunneo-pubescentibus, utrisque antice fasciculo parvo pilorum nigrorum fulvorumque; plaga laterali brevi triangulari, omnino ante medium, plaga submediana communi sub-rhomboidali. Long.* 25-29 *mm.*

Carin (district of Chebà), alt. 900-1100 m.

This species is somewhat larger than the preceding, and resembles it in general facies and coloration. The elytra have a brownish pubescence, with four fulvous-brown plagae — one basal, triangular; one triangular, on each side, and placed wholly in front of the middle; the fourth common, for the most part behind the middle, and somewhat roughly rhomboidal in form. The anterior border of this plaga and the posterior borders of the lateral plagae are somewhat mixed with grey. At the base of each elytron, just about the middle of the lateral border of the basal triangular plaga, there is a fascicle of black and fulvous hairs.

151. **Niphona ornata**, sp. n. *Pube ochraceo-alba dense vestita, elytris basi supra humeros et plaga magna angulata dorsali fuscis; prothorace supra sparsim punctato, medio breviter carinato, lateribus utrisque antice uni-dentato; elytris aculeatis, basi utrinque fasciculato-cristatis; apicibus oblique truncatis sat dense irregulariterque fimbriatis; antennis quam corpore vix longioribus (♂); paullo brevioribus (♀), subtus ciliatis, articulis 5° ad 11um, et apice quarti, fuscis; tarsis nigris. Long.* 12$^1/_2$-16 *mm.*

Carin Mts. (Chebà district), alt. 900-1100 m.

Clothed with a dense pubescence which differs in shade on different parts of the body according to the predominance of the white or of the fulvous-brown hairs of which it is mostly composed. The head and basal joints of the antennae are of a pale fulvous-brown colour, the prothorax, patches on the sides of the elytra, the underside of the body and of the legs are of a more or less whitish colour. Between the basal crest and the shoulder of each elytron there is a dark fulvous brown patch which is interrupted in the middle by a whitish line. A large angular dark-brown plaga occupies a little more than the median third of the disk. Its anterior borders — directed obliquely backwards on each side from the suture — and its outer borders are sharply definel; while posteriorly it passes indefinitely into the pale fulvous-brown of the hinder part of the elytra. Prothorax with a rather strongly marked tooth placed anteriorly and low down on each side; the disk with a short median carina which is very narrow anteriorly, but which slightly widens and bears a feeble groove posteriorly; the anterior border of the pronotum bears a series of very small fuscous spots. The elytra, broad at the base, are much narrowed from thence to the apex, where they are each somewhat obliquely truncate and fringed with rather long hairs.

In general shape and in coloration this species most nearly resembles *N. plagiata*, White; but the prothorax of the latter is quite rough, and does not bear a tooth on each side. Its eyes also are a little larger.

152. **Niphona parallela**, White, var. — *N. parallela*, White, Ann. and Mag. Nat. Hist., Ser. 3, Vol. II, p. 267.

Tenasserim, Thagatà and Upper Burma, Bhamò; also China and North India (Brit. Mus. collection).

The single male example which served as the type of *Niphona parallela*, has a broad fuscous band which extends along the middle of each elytron from the base to beyond the middle. This band is wanting in the present variety, in which the elytra have everywhere a greyish pubescence mixed with fulvous brown. This is the only difference which I can detect. The eyes in this species are rather large; and the second abdominal ventral segment of the male has on each side a transverse semi-oval depression lined with greyish or fulvous hairs, similar to, but rather larger than, the hairy depression to be found in the same position in the males of *Niphona picticornis*, Muls.

153. **Camptocnema lateralis**, White. — *Nyphona lateralis*, White, l. c., p. 267. — *Mylothris bimaculata*, Brong., N. Archives du Muséum, Ser. 3, Vol. III, p. 267, pl. 10, fig. 11.

Carin Chebà; alt. 900-1100 m.; also occurs in Siam and Assam.

154. **Pterolophia lateralis**, sp. n. *Nigro-fusca, pube fulvo-brunnea obtecta: pronoto longitudinaliter cinereo fuscoque vittato, elytris utrisque ad medium lateris plaga fusca supra convexa et late cinereo-limbata; antennis articulo 1° griseo-fulvo, 4° fere toto cinereo, ceteris fuscis obscure griseo-annulatis. Long.* 10-11 *mm.*

Pronotum with alternating longitudinal bands of darker and lighter coloured pubescence, with two darker bands on each side of the middle and separated by a faint greyish line, these are succeeded on the outside by two pale fulvous or ashy lines — one on each side, and external to these is a darker vitta on each side. The elytra are strongly enough narrowed from the base, appearing as if compressed from side to side; posteriorly they are strongly declivous; the apices are obliquely truncate and very faintly emarginated towards the suture; on each elytron there is a distinct lateral median fuscous plaga, which, towards the upper side, has a convex margin and is surrounded by a rather broad arcuate band of an ashy-white or fulvous-white colour. The posterior declivous portion of the elytra is also in the middle more or less ashy or pale-fulvous in colour. The punctuation of the elytra is rather strong, and is arranged somewhat in rows, which on the middle of the disk are separated by slightly raised lines.

This species may be recognized by the absence of a basal tuft or tubercle from the elytra, taken with its rather distinct style of marking.

As it seems impossible to fix upon any definite characters which separate *Praonetha* from *Pterolophia*, I have been obliged to treat the first of these names as a synonym.

155. **Pterolophia modesta**, sp. n. *Brunneo-pubescens, fusco punctata et maculata; capite fulvo-brunneo supra maculis duabus parvis fuscis; prothorace brunnescente, sparsim fusco-punctato, supra bi-tuberculato, plaga media prope basim pallide fulva; scutello pallide fulvo, macula media fusca; elytris brunnescentibus, fusco-punctatis utrinque pone medium fulvo plagiatis, utrisque tuberculo parvo basali et cristis duabus fasciculatis, una ante, altera, pone medium; antennis (♂) corpore aequalibus. Long.* 11-14 *mm.*

Carin Mts. (district of Chebá) and Tenasserim (valley of Houngdarau).

Clothed with a brownish or fulvous-brown pubescence, with a few paler patches, and with many small points and a few spots of a dark velvety brown colour. Head moderately concave between the antennal tubercles. Prothorax transverse, sides nearly straight and very sligthly diverging from the base to within a short distance of the anterior margin where they become constricted; the disk transversely sinuately impressed near the apex, with a fine and straight transverse groove just before the base, and with a broad and rather faint depression along the middle; on either side of this depression there is a small tubercle. The elytra have each a small tubercle at the base, and, immediately behind the tubercle, a short longitudinal crest, while further back is another longitudinal crest which ends just at the beginning of the posterior declivity. The tubercle and crests carry tufts of hairs; the anterior crest having a continuous tuft, while on the posterior there are small tufts set at intervals. External to the crests there are on each elytron one or two raised lines. The punctuation of the elytra is rather strong and dense towards the sides, but appears finer and less dense on the disk towards the apex. Last ventral segment of the abdomen dark brown, hind margin of the first segment with a fringe of pale fulvous hairs; intermediate segments spotted with pale fulvous and dark brown.

156. **Pterolophia subfasciata**, sp. n. *Capite prothoraceque fulvo-pubescentibus; elytris subelongatis dense fortiterque punctatis, pube fulvo-brunnea obtectis, fascia transversa irregulari ad medium fulvo-nigra; corpore subtus pedibusque piceis, griseo-pubescentibus; antennis utroque sexu quam corpore paullo brevioribus, articulis 1°, 2^oque griseis, ceteris apice fuscis, basi cinereis. Long. $8^1/_2$-13 mm.*

Carin Mts. (Chebà district), alt. 900-1100 m.

Head and prothorax with a not very dense fulvous pubescence. The prothorax closely punctulate, and with two somewhat obsolete fuscous vittae along the middle of the disk. Elytra elongated for this genus, almost uniformly convex above, somewhat gradually declivous posteriorly; traces of costae on the disk are scarcely apparent; the pubescence is rather close and varies from almost a grey to a reddish-brown colour, it is scantier over the irregular and not very distinct band which thus appears darker than the rest of the surface; on this band also, especially at the sides, the close and strong punctuation of the elytra is more distinctly seen, while over the rest of the surface owing to the closer pubescence, it appears feebler and less dense. The apices of the elytra are very slightly truncate near the suture or are entirely rounded.

157. **Pterolophia proxima**, sp. n. *Precedenti persimilis sed differt elytris pone basim transversim depressis, postice valde declivis, an-*

tennis fere omnino griseis, articulis 5° ad 9um apice extus dentatis. Long. 8 $^1/_2$ mm.

Carin Mts. (Chebà' district). One example.

The coloration of this species very much resembles that of the preceding; but the antennae are almost entirely fulvous-grey: these also have a tooth at the outer apex of each of the joints from the third to the ninth. The elytra are somewhat shorter than in the last species, are transversely depressed a little behind the base, and are strongly and somewhat abruptly declivous behind.

158. **Pterolophia consularis**, Pasc. — *Praonetha consularis*, Pasc. Proc. Zool. Soc., 1866, p. 240.

Carin Mts. (districts of Chebà and Ghecù), alt. 900-1400 m.

The examples taken agree with the type in all respects, except that the sides of the prothorax are not so dark-brown in colour.

A single example obtained at Metanja in Burma, differs from the Carin specimens in having the prothorax uniformly greyish brown above, and without the paler patch along the middle of the disk. This specimen may be considered as belonging to a variety.

159. **Pterolophia armata**, sp. n. *Pube brevi fulvo-grisea sat dense obtecta; prothoracis dorso utrinque vitta lata brunnescente. Elytris brunnescentibus, vitta suturali ante medium abbreviata, et fascia obliqua utrinque pone humerum pallidioribus, utrisque macula parva ad basin, macula parva supra tuberculum posita paullo pone basin, et maculis tribus vel quatuor ad summum declivitatis posticae, nigris; coxis anticis utrisque (♂) tuberculo parvo acuto armatis; tarsis plus minusve nigro-fuscis; antennis fulvo-brunnescentibus, articulis 5°, 6°que nigris. Long. 11.5-12.5 mm.*

Carin Mts. (district of Chebà), alt. 900-1100 m

Body underneath, last ventral segment which is black excepted, legs, sides of prothorax and front of head clothed with a pale fulvous-grey pubescence; body above and elytra brownish, with a paler fulvous-grey band extending along the middle from the head to a little beyond the anterior third of the elytra;

and with an oblique greyish patch on the side of each elytron just below and behind the shoulder, each elytron also with some small black spots of which one is placed at the extreme base, a second on a tubercle a little behind the base, and three or four forming a transverse band at the beginning of the posterior declivity. Three or four longitudinal costae may be distinguished on the disk of each elytron. The antennae are fulvous-brown, with the fifth and sixth joints blackish; these two joints are together scarcely as long as the fourth, the latter is slightly curved and is appreciably longer than the third joint. The anterior coxae are each armed a small sharp tubercle which is directed towards the middle line.

160. **Pterolophia annulata**, Chevr. — *Coptops annulata*, Chevr. Revue Zoologique, 1845, p. 99.

Upper Burma (Shwegoo, Metanja) and Carin Mts. (Chebà district).

The greyish-white patches on the elytra are more limited in extent than in the type, and in this respect resemble examples from North India. The specimens from China and Hong-Kong in the British Museum collection show variation in the amount of grey coloration of the elytra; so that the difference just mentioned cannot be considered of any great importance.

161. **Pterolophia nigrocincta**, sp. n. *Modice elongata, sub-cylindrica, capite prothoraceque griseo-pubescentibus maculis minutis nigris adspersis; elytris flavescente-pubescentibus, fascia lata transversa ad medium, maculis irregularibus versus basin, et punctis postice nigris; antennis articulis 1° ad 4um griseo subtiliter pubescentibus, ceteris nigris; corpore subtus pedibusque griseis. Long. 8-9 mm. Lat. $2^1/_2$-3 mm.*

Carin Mts. (Chebà district).

This is a rather narrow and somewhat elongated species. The prothorax has its sides nearly parallel. The elytra are almost regularly convex above, gradually declivous behind, with their sides almost parallel in their anterior two-thirds, they are rather closely punctured; have a yellowish pubescence, with a well marked broad dull black band crossing them at the middle, and with two black patches at the base — one on each side — which unite together or with a smaller third spot, at a short distance behind the scutellum. Between these basal patches and the median band as well as towards the apex there are numerous minute black points.

162. **Pterolophia socia**, sp. n. *Precedenti similis, sed fascia transversa elytrorum angustiori, prothorace supra uniformiter pubescente, haud maculis minutis nigris interrupto. Elytris pube flavo-grisea haud nigro-punctata; antennis totis griseo-pubescentibus. Long.* 7 $^1/_2$-9 *mm.*

This species much resembles the last; but the greyish or yellowish-grey pubescence of the prothorax is not interrupted by minute black spots; the transverse black fascia at the middle of the elytra is much narrower; the yellowish-grey pubescence between the basal spots and the median band and between the latter and the apex is not sprinkled with small black points. The antennae are entirely covered with a greyish pubescence.

163. **Pterolophia alboplagiata**, sp. n. (Pl. I, fig. 12). *Parva, fusca; pube fulvo-brunnea vestita, prothoracis lateribus, et fascia transversa ad medium elytrorum, albo-cinereis; prothorace crebre elytrisque dense et fortiter punctatis; his utrisque prope basin tuberculo nigro-piloso instructis. Long* 7 *mm.*

Carin Mts. (district of Chebà).

Prothorax closely **punctured**; **with a** pubescence which **is** fulvous-brown along the middle of the disk, and ashy-white at **the sides. Elytra with their sides sub-parallel** for the anterior two-thirds, thence narrowed to the apex; very distinctly, strongly, and rather closely punctured; each with a well marked centro-**basal tubercle surmounted by a** fascicle of black hairs; with a **transverse whitish** fascia which reaches from the outer margin **to near the suture and** is much narrowed at its inner extremity; **with some** small and irregular spots of whitish pubescence near **the base and apex. The legs** have an ashy-grey pubescence. The antennae are a little shorter than the body, with the last three joints, and the apices of joints 3, 4, 5 and 8 fuscous, the remainder fulvous or cinereous.

164. **Pterolophia persimilis**, sp. n. *Precedenti persimilis, sed paullo minor, prothorace densius punctulato, ad latera haud cinereo-plagiato, fascia sub-mediana elytrorum sutura haud attingente. Long.* 5.5-6.5 *mm.*

Bhamò in Upper Burma. One example. Also occurs in the neighborhood of Hong-Kong (*Bowring*).

This species strongly resembles *P. alboplagiata*, but the prothorax appears more thickly punctured (though this may be due to its somewhat scantier pubescence), and is not ashy-white at the sides. The white band, or rather plaga, near the middle of each elytron does not reach to the suture. The legs are dark brown with a thin, and in some parts, scattered pubescence of a greyish colour.

165. **Pterolophia fulvisparsa**, sp. n. *Parva, nigro-fusca, dense fortiterque punctata, pube fulvina dispersa; elytris utrisque prope basin tuberculo parvo nigro-piloso; antennis quam corpore vix brevioribus, fere toto fuscis; pedibus fuscis, fulvo annulatis. Long.* 5 $^{1}/_{2}$ *mm.*

Thagatà. One example.

Blackish brown, with a scattered fulvous pubescence. Prothorax very thickly and less strongly punctured; elytra thickly and very strongly punctured, each near the base with a small tubercle surmounted by a short tuft of black hairs. Legs with the bases of the femora and apices of tibiae fuscous, the intermediate parts testaceous with a fulvous pubescence; the tarsi more or less piceous.

166. **Pterolophia quadrifasciata**, sp. n. *Pube fulvo-brunnea, fulvo-griseoque mixta, sat dense vestita; pronoto medio vittis obsoletis et elytris plagis duabus utrinque albo-cinereis; prothorace lateraliter sub-rotundatus; elytris sat brevibus postice ad apicem valde angustatis, basi utrinque tuberculo nigro-fulvoque piloso, pone medium utrinque tricostato, costa interna quam externis breviori sed distinctiore; antennis quam corpore brevioribus, articulis* 1°, 2° *fulvobrunneis,* 4° *fulvo-cinereoque; ceteris fuscis cinereo annulatis. Long.* 9-10 *mm.*

Thagatà in Tenasserim; and Carin Mts. (Chebà district).

Elytra each with a distinct tubercle at the base, the tubercle having blackish brown hairs at its summit. About three costae are with difficulty to be distinguished on the intermediate third of each elytron, the innermost costa only, which is the shortest being at all distinct. On the disk of each elytron there is a small greyish area from which two ashy-white bands or patches pass off — one running obliquely outwards to reach the margin just below the shoulder, the other at first passing slightly backwards runs then in a nearly transverse direction to the outer margin.

This species has some resemblance, in its mixed pubescence and general outline, to *P. annulata*, Chevr. and *P. scopulifera*, Pasc.; but may be easily distinguished by the very distinct tufted tubercle or crest at the base of each elytron.

167. **Pterolophia carinata**, sp. n. *Pube brunnea, fulvo-griseoque mixta sat dense vestita; prothoracis lateribus sub-parallelis, dorso fere impunctato; elytris dense fortiterque punctatis, utrisque basi tuberculo cariniformi et pone medium carina brevi valde prominenti, lineaque elevata, apicibus oblique subsinuato-truncatis; antennis corpore fere aequalibus, brunnescentibus, articulis apice anguste griseis. Long.* 10 mm.

Shwegoo in Upper Burma. One example.

Clothed with a pubescence of somewhat mixed colours, the prevailing tint being a light chocolate brown. The disk and the posterior declivous portion of the elytra are more greyish or fulvous. The prothorax is entirely impunctate, unless some dark brown minute rounded spots mark out the position of some feeble punctures. The elytra are thickly and strongly punctured; each has a basal cariniform tubercle, a short prominent carina extending along the disk in the intermediate third of the length of the elytra, and ending just where the posterior declivity begins, and a slightly raised line external to this carina, extending a little beyond it behind, while in front its course may be traced up to the base. Between the basal tubercle and the anterior extremity of the carina which lies nearly in the same line with it, there is a shallow and somewhat oblique depression on each elytron.

This species appears to be somewhat nearly allied to *Pterolophia camura*, Newm.; but has a generally paler and more uniform coloration, a more distinct dorsal carina on each elytron, and a more parallel-sided prothorax. The disk of the prothorax also is less convex and nearly flat; the apices of the elytra are sinuately truncate, while in *P. camura* they are almost rounded.

168. **Pterolophia vagans**, sp. n. *Fulvo-pubescens; prothorace transverso, sat dense punctulato, lateraliter vix rotundato, supra medio vitta, male limitata, grisea. Elytris sat dense, haud fortiter punctatis, postice declivis, lineis duabus vel tribus, paullulo elevatis et fere obsoletis utrinque instructis, pube fulva, plaga suturali ante- et fascia transversa angusta, paullo pone medium, cinerascentibus; apicibus rotundatis; antennis ($\female$) medium elytrorum paullo superantibus, articulis 1° 3° et 4° subaequalibus, 5° quam 4° multo breviori, 5°-11um longitudine gradatim decrescentibus; scapo apice cicatricoso, cicatrice levi, sub-nitida, acute marginata. Long. 14 mm.*

Carin Mts. (district of Chebà), alt. 900-1100 m. One female example.

By adhering strictly to the system followed by Lacordaire, a new genus, in the group of the *Mesosides*, should be formed for the reception of this species. Though it differs from all the other species of *Pterolophia*, and indeed from all the *Niphonides*, in having a smooth cicatrix, limited by a sharp edge, on the apex of the antennal scape, yet the general ensemble of its characters is such as to bring it into very close affinity with certain forms of *Pterolophia*. It might be right perhaps to regard it as a representative of a new genus which brings into greater prominence the undoubted affinities that exist between the *Mesosides* and the *Niphonides*. It seems to me to be one of the faults of Lacordaire's arrangement that these two groups are placed so far apart. And I also think that Lacordaire's sub-division of the latter group may be improved upon by taking into consideration the structure of the mesonotum. In the genera more nearly allied to the *Mesosides* the mesonotum is pointed in front and is furnished with a stridulating surface; while in the majority of the genera (including *Niphona* and closely related forms) the mesonotum is not so prolonged and pointed in the middle, and does not, so far as I can find, possess a stridulating surface.

169. **Pterolophia scopulifera**, Pasc. — *Praonetha scopulifera*, Pasc., Trans. Ent. Soc. Lond. 1865, p. 175.

Bhamò, in Upper Burma.

170. **Lychrosis zebrinus**, Pasc. — *Hathlia zebrina*, Pascoe, Trans. Ent. Soc. Lond., Ser. 2, Vol. IV, p. 252.

Carin Mts. (district of Chebà), alt. 900-1100 m.

171. **Lychrosis humerosus**, Thoms. — *Mycerinus humerosus*, Thoms., Syst. Ceramb., App. p. 550.

Taken at Teinzò and Bhamò in Upper Burma.

172. **Lychrosis** (?) **angustus**, sp. n. *Griseo sat dense pubescens; prothorace vittis duabus dorsalibus elytrisque plaga laterali utrinque pone medium pallidioribus; elytris punctis nigris dispersis; prothorace cylindrico quam latiori vix longiori, dense minuteque punctulato; elytris elongatis, dense punctatis, apicibus emarginatotruncatis; antennis (♂) quam corpore paullo longioribus, articulo 4° quam 3° longiori, sequentibus utrisque quam 3° multo brevioribus; antennis (♀) quam corpore brevioribus. Long.* 8-10. *Lat.* 2-2.5 *mm.*

Hab. Carin Mts. (district of Chebà), alt. 900-1100 m.

In its narrower and more elongate form, this species departs from the more typical members of the genus; the antennae also differ in having the fourth joint longer than the third. But beyond these differences there seems to be no good character by which the species could be separated generically from *Lychrosis*.

173. **Lychrosis ?** sp. n.

Carin Mts. (Chebà district).

174. **Sthenias Pascoei**, Rits., Notes Leyd. Mus., vol. X, p. 272. — *Sthenias grisator*, Pasc., Trans. Ent. Soc., Ser. 3, vol. III, p. 160 (nec Fabr.).

Carin (district of Chebà); occurs also in Sumatra, Java, and in North India.

175. **Apomecyna leucosticta**, Hope. — *Callidium leucostictum*, Hope, in Gray's Zool. Miscellany (1831), p. 28.
Carin Mts. (district of Chebà), alt. 900-1100 m. Occurs also in Nepal, North India.

176. **Apomecyna histrio**, Fabr. — *Lamia histrio*, Fabr., Ent. Syst., I, 2, p. 288.
Carin Mts. (district of Chèbà), alt. 900-1100 m.

177. **Apomecyna pertigera**, Thoms., Physis, II, p. 160.
Palon in Pegu. One example.

178. **Apomecyna cretacea**, Hope. — *Callidium cretaceum*, Hope, in Gray's Zool. Miscellany, p. 28.
Carin Mts. (district of Chebà).

179. **Mycerinopsis lineatus**, sp. n. *Elongatus, pube griseo-flava obtectus; capite inter tuberculos antenniferos fortiter concavo, fronte sparsim, vertice densius, punctato; prothorace quam latitudine vix longiori, distincte sat sparsim punctato, lateribus fere parallelis; elytris dense punctatis, griseo-flavo-pubescentibus, pube in lineis quatuor utrinque paullo elevatis et impunctatis condensata; apicibus utrisque sub-attenuatis et fere rotundatis; antennis (σ) quam corpore fere sesqui-longioribus, griseo leviter pubescentibus. Long.* 11-17 *mm.*

Carin Mts. (Chebà district), 900-1100 m.; and Thagatà in Tenasserim.

The pubescence which is of a yellowish or yellowish-grey tint above and grey on the underside, is somewhat uniformly spread over the whole body, but is slightly denser along certain lines of which one passes along the middle of the pronotum and four slightly raised along each elytron; of these four lines the two inner extend almost from the base and unite together at a short distance from the apex, and are thence continued back as a single line to meet another short line which arises from the junction of the two outer lines; the outermost line may be traced forwards almost up to the shoulder, the one next it disappears anteriorly after passing the middle of the elytron. The punctures of the elytra are, behind the middle, arranged more or less regularly in double rows between the raised lines; anteriorly their numbers increase and the regularity of their arrangement is disturbed.

The structural characters by which this species differs from *M. aridus*, Pasc. — the type of the genus — are not sufficient to justify the creation of a new genus for its reception. With the exception of its more parallel-sided prothorax, its slightly longer antennae, and more prominent antennal tubercles, it agrees essentially in generic detail with *M. aridus*.

180. **Eunidia simplex**, Gahan, Ann. and Mag. Nat. Hist., ser. 6, vol. V, p. 64.

Carin (district of Chebá). One example.

181. **Ropica**, sp.

Thigyam in Burma. One example.

182. **Atimura terminata**, Pasc., Trans. Ent. Soc., ser. 3, vol. I, p. 548, pl. 23, fig. 6.

Rangoon. One example.

This example does not seem to be specifically distinct from the type of the above species, with which I have compared it. The latter came from Port Denison in Queensland.

183. **Atimura apicalis**, sp. n. *Fusca, cinereo-pubescens; capite antice et supra dense fulvo-ochraceo-pubescente; elytris apice abrupte fortiterque declivis, declivitate luteo-albido dense pubescente et fasciculata. Long. 8 mm.*

Hab. Carin Mts. (district of Chebá), alt. 900-1100 m. One example.

This species may be easily distinguished from the preceding by the more abrupt manner in which the elytra are turned down at the apex. This apical declivous portion is closely covered with a dirty-white pubescence, and is also furnished with some small tufts of a similar colour — about four to each elytron, of which one is at the **summit of** the declivity and marks the posterior extremity of a very short and feebly raised carina which is in a line with some granules placed at wide intervals along the disk of the elytron; the second tuft is at about the middle of the declivity, and the remaining two are at the posterior border and give to the latter a deeply emarginate appearance. The prothorax is rather feebly and not very closely punctured, it is slightly rugose towards each side of the disk, and on the middle of the disk bears four minute granule-like tubercles. The elytra are strongly and closely punctured, the punctures being only partially hidden by the not very dense pale-grey pubescence. The antennae are about equal in length to the body. The single specimen taken is probably a female.

184. **Sybra (?) posticata**, sp. n. *Flavo vel fulvo-albido-pubescens, capitis fronte, prothoracis disco elytrisque plus minusve brunnescentibus; capite inter tuberculos antenniferos sat anguste fortiterque concavo, occipite maculis duabus fusco-velutinis; prothorace supra ante medium longitudinaliter rugoso, postice rugoso-punctato; scutello albido; elytris postice sat valde declivis, parte declive pallidiore, apicibus truncatis; femoribus intermediis posticisque versus apicem et abdominis lateribus fusco irregulariter denseque maculatis; segmento ultimo medio fusco-plagiato; antennis (σ) corpori aequalibus, (φ) brevioribus: griseis, articulis intermediis infuscatis.* Long. 11-13 mm.

Catcin Cauri and Bhamò in Upper Burma. Thagatà in Tenasserim; also Darjeeling and Cambodia (Brit. Mus. coll.).

The antennal tubercles are somewhat more prominent on the inner side and separated by a narrower interval (somewhat v-shaped when looked at from above) than is usual in the genus *Sybra*. The elytra also are more strongly declivous posteriorly. Each of the intermediate cotyloid cavities, though in fresh specimens appearing to be completely shut off from the epimeron on the outside, in reality extends out to it; the narrow outward extension being occupied by the trochantin when the femur is turned backwards. Lacordaire has laid stress upon the closing in of the intermediate cotyloid cavities in the group to which *Sybra* belongs. « Pris dans son ensemble, il est très-voisin des Apomécynides et n'en diffère essentiellement que par la fermeture des cavités cotyloïdes intermediaires ». This is not however, a reliable character. A careful examination of the different species of the *Ptericoptides* will show that in very few, if any, are the intermediate cotyloid cavities completely shut off from the mesothoracic epimeron. They often, as in the species just described, appear to be so, owing to the concealment of the narrow outer prolongation of the coxal cavity by the rather thick pubescence which overlaps it, the trochantin itself also being often slightly pubescent on its outer edge. In many of the species of the *Ptericoptides* that I have examined the mesothoracic epimeron is more distinctly in connection with the cotyloid cavity than is the case in such genera as *Apomecyna* and *Mycerinopsis*.

185. **Sybra procera**, Pasc. — *Hathlia procera*, Pasc., Trans. Ent. Soc., ser. 2, vol. V, p. 50. — *Ropica cylindrica*, Pasc., Trans. Ent. Soc., 1888, p. 504.

Palon in Pegu, and Carin (district of Chebà). Occurs also in India and Ceylon.

186. **Pothyne variegata**, Thoms.? Systema Ceramb., p. 97; Lacord., Gen. des Coleopt., IX, 2, p. 694.

One specimen taken at Teinzò in Upper Burma. Also occurs in Assam and Siam.

The descriptions given by Thomson and Lacordaire (and probably drawn up from rubbed specimens) do not quite accurately fit the examples now before me, so that I have some doubt in referring the latter to Thomson's species.

The pubescence is grey, with seven fulvous-red lines along the thorax and the same number on the elytra. The latter in fresh specimens have numerous small dull black spots, and two larger sub-nitid black spots — one on each shoulder. Of the seven reddish lines on the elytra, one is a common sutural line, one at the extreme outer margin of each elytron; the three intermediate lines on each side are here and there interrupted in their course. The front of the head is rather thickly and somewhat asperately punctured, the prothorax is obsoletely punctured above, and is rather feebly wrinkled across the middle. The elytra are thickly but not very strongly punctured. The vertex of the head in this, as in almost every species of the group, has two short black glabrous and slightly elevated lines, placed one on each side of the middle line and extending forwards from the occiput.

187. **Hyllisia lineata**, sp. n. *Fusca; capitis fronte verticeque dense punctatis; prothorace sat dense punctato, vittis quinque pallide fulvis; elytris dense punctulatis, fusco-castaneis, sutura et utrinque lineis quatuor vel quinque pallide fulvis, apicibus sub-truncatis; corpore subtus pedibusque subtiliter griseo pubescentibus, pectore vitta fulva utrinque; antennis (♂) quam corpore plus duplo longioribus, subtus usque ad medium longe sat denseque ciliatis, articulo 1° basin prothoracis fere attingente. Long.* 15 *mm.*

Teinzó in Upper Burma; also India (Brit. Mus. coll.).

The antennae in the male of this species are more than twice as long as the body, the first joint, cylindrical in form or very slightly flattened underneath, reaches almost to the base of the prothorax, it is ciliate underneath; the third joint, which is distinctly longer than the first is, together with the following joints up to about the middle of the sixth, provided with a thin fringe of very long hairs underneath. The head is thickly punctured in front, as well as on the vertex. The prothorax is strongly enough but not very thickly punctured. It has three narrow pale-fulvous vittae above and one — broader — on each side. The elytra are thickly but rather feebly punctured; the suture and four or five lines on each elytron are covered with a pale-fulvous pubescence, the intervals almost, glabrous and subnitid, are of a dark chestnut-brown colour. The legs and underside of the body have a slight greyish pubescence, with a pale fulvous vitta on each side of the meso and metathorax. The hind femora reach to about the middle of the second-abdominal segment.

188. **Hyllisia consimilis**, sp. n. H. lineatae *persimilis sed minor; scapo antennarum apice anguste cicatricoso; apicibus elytrorum oblique truncatis. Long.* 10-11 *mm.*

Carin Mts. (Chebá district), alt. 900-1100 m.

This very closely resembles the preceding species, but is distinctly smaller; the head is less strongly and less thickly punctured above and exhibits no trace of the two feebly elevated glabrous lines; the delicate hair fringe of the antennae is not so long (this character may not be of specific importance, as in one or two undescribed **species the** individuals seem to vary **with regard to the length of the hairs or ciliae on** the antennae); **the scape of the antennae** has, at the inner apex, a narrow **cicatrice** bounded **by a** very fine edge. There is some trace of **a** cicatrice in the same position in *H. lineata*, but in this **species I have** not been able to detect a limiting carina. In the **present species also the apices of** the elytra are somewhat more **distinctly truncate.** Closely as this species resembles *H. lineata*, its **resemblance to a** South African form (? *H. subvirgata*, Fairm.) is scarcely less pronounced. The latter species has the same **general facies and style of coloration.** The scape **of its** antennae **is also** furnished with a very narrow smooth cicatrice, **limited by a sharp but very feeble** carina. The apices **of the elytra** are more obliquely truncate, becoming in fact sub-acuminate. The most essential difference between the two species lies in **the prosternum, which** in **the African** species is **not** only narrower between the coxae, but does not widen **out to nearly so great** an extent posteriorly.

Hippopsis tonkinea, Fairm., appears **to be very much like** *Hyllisia consimilis* and probably belongs to the **same** genus.

189. Smermus sericeus, sp. n. *Fuscus, pube griseo-sericea, elytris plagiatim condensata, vestitus; prothorace vittis tribus fulvo-griseo-sericeis; capite fronte sat dense, vertice densissime punctato, prothorace elytrisque sat dense punctatis; his punctis paullo pone medium evanescentibus, apicibus utrisque fere regulariter rotundatis, corpore subtus pedibusque pube fulvo-grisea sericea sat uniformiter* **denseque** *vestitis. Long.* 15-17 *mm.*

Carin Mts. (Chebá district), alt. 300-1000 m.; also Burma (Brit. Mus. coll.).

Though **at first sight looking** rather unlike *Smermus Mniszechi*, Lac. — **the type** of **the genus** — there can be little doubt that the true place of the present species is in this genus, with which it agrees in all its chief structural characters. The prothorax **is somewhat** shorter and is more thickly punctured than in *Mniszechi*. The elytra are each almost regularly **rounded at the apex. The length of the antennae** in the examples before me is from three to four times that of the body.

A closely allied species occurs in Java. This species (*Smermus similis*, sp. n.) differs from *sericeus* in having slightly obliquely truncated apices to the elytra. The size is smaller. Long. 12 mm.

The pale fulvous-grey silky pubescence of the elytra forms somewhat irregular longitudinal bands uniting at the apex, and with a darker, somewhat brownish, pubescence filling in the intervals anteriorly. The vertex of the head is much less thickly punctured.

190. **Tetraglenes bucculenta**, sp. n. *Linearis, fusco-testacea, pube fulvo-brunnea sat dense obtecta; capitis vertice prothoracisque dorso linea media, et utrinque vitta minus dense pubescentibus; capite basi lato, genis prominentibus; elytris crebre sat fortiter punctatis, lateribus fere parallelis, versus apicem angustatis, apicibus oblique truncatis, sub-divaricatis; antennis (♀) corpore aequalibus, fuscis; subtus longe, supra minus longe, sat denseque ciliatis. Long.* 12 mm.

Teinzò in Upper Burma. One female example.

This species ought, perhaps, to constitute a genus distinct from *Tetraglenes*. Its head is somewhat differently shaped, being very much wider at the base, with prominent cheeks, and (considering the size of the insect) very strong mandibles; the antennae in the female are quite as long as, or rather longer than, the body; the relative proportions of the joints are however nearly the same as in *T. insignis*, Newm., with the exception of the first joint which is relatively shorter. The elytra have their sides nearly parallel in their anterior fourth fifths, and are narrowed towards the apex, where each is cut back obliquely from the suture, so that they appear slightly divaricate.

The species, though more uniform in coloration, bears a strong general resemblance to *Tetraglenes insignis*, and may provisionally be considered congeneric with it.

191. **Tetraglenes insignis**, Newm., The Entomologist, vol. 1, p. 300.

Bhamò in Upper Burma.

192. **Eucomatocera vittata**, White, Ann. and Mag. Nat. Hist., XVIII (1846), p. 49, pl. I, fig. 4.

Carin Mts. (district of Chebá), alt. 900-1100 m.; Teinzó in Upper Burma.

The head in this species is shaped like that of *Tetraglenes insignis*; and the eyes also offer the same characters. White unfortunately and, we may add, carelessly described and figured a specimen in which the head was misplaced. The front of the head was uppermost, the vertex in front, while that part of the head which ought to have been inserted in the prothorax did duty for a mouth. No wonder, therefore, that, as White remarks, the mouth-parts were wanting. This explanation will account for Lacordaire's observations on the genus.

Estigmenida, gen. nov.

Intermediate cotyloid cavities open. Claws of tarsi divergent. A sinus on the intermediate tibiae. Head with the front trapeziform and oblique, the antennal tubercles somewhat prominent and slightly divergent; the mouth parts applied to the anterior coxae when the head is fully retracted. Prothorax slightly convex above, scarcely longer than broad, with the sides sub-parallel and unarmed; with the pronotum in the middle line about twice as long as the prosternum; the latter forming a very narrow boundary to the coxal cavities in front, rather narrow in the middle between the coxae, but widening out considerably behind. Elytra nearly one half broader than the prothorax, their sides very gradually narrowed and almost parallel in their anterior three fourths and thence more strongly narrowed to the apex; the latter truncate. Each elytra with three or four costae of which one lateral and outermost is somewhat more prominent than the rest. The mesosternum is, like the prosternum, almost flat; it gradually narrows posteriorly. The legs are of moderate length; the femora slightly incrassate below the middle; the middle coxae of the male each with a small sharp conical tubercle at its inner summit.

Antennae of the male about equal in length to the body; first joint rather short and thick, the third nearly twice as long, the fourth distinctly shorter than the third, the fifth and following subequal among themselves, each very much shorter than the fourth; the third, fourth and fifth with a short thick fringe of hairs underneath.

The antennae of the female resemble those of the male; but are much shorter; surpassing by a little only the middle of the elytra.

The combination of characters above described shows that this genus does not fit well into any of Lacordaire's groups. It is undoubtedly nearly allied to *Pemptolasius*, to which I have already, with some doubt, assigned a position near *Ectatosia*.

193. **Estigmenida variabilis**, sp. n. (Pl. I, fig. 13). *Rufo-castanea, fere glabra, sat nitida; capite modice punctato, vertice genisque albo-cinereo-vittato; prothorace grosse sat denseque punctato, supra utrinque obsolete cinereo-vittato; elytris dense fortiterque punctatis, utrisque lineis quatuor elevatis, quarum duabus distinctioribus; antennis articulis 3° ad 5um nigro-fuscis et subtus breviter denseque nigro-fimbriatis, articulis 6° ad 11um tenuibus, plus minusve pallide-griseis. Long.* 11-13 mm.

Var. A. *Corpore toto elytrisque nigris.*

Var. B. *Corpore rufo-testaceo; elytris pedibusque flavo-testaceis.*

Carin Mts. (Chebà district); alt. 900-1100 m.

The resemblance of this species to *Estigmena chinensis*, Hope — a tolerably common Oriental Hispid — is most striking, and appears to me to be the result of something more than mere accidental causes. Not only are the colours, punctuation, and whole general aspect of two species very much alike; but the antennae of the Longicorn do not show so great a difference from those of the Hispid, as, from their greater length, they might be expected to. This is due to the fact that the last six joints are slender, are mostly of a pale-greyish colour, and are not fringed with hairs like the basal joints, and in consequence are not so distinctly visible at a short distance. What advantage the Longicorn could derive from thus mimicking the Hispid is not obvious; indeed the sober style of coloration, or absence of bright warning colours, seems opposed to the idea that the Hispid might be recognized by insectivorous animals as a distasteful morsel.

194. **Pemptolasius humeralis**, Gahan, Ann. and Mag. Nat. Hist. ser. 6, vol. V, p. 65, pl. VII, fig. 8.

Carin Mts. (district of Chebà); also occurs in North India.

195. **Rondibilis plagiata**, sp. n. *Sparse setosa; griseo-cinereopubescens; elytris plaga basali communi, triangulari vel v-formante, et utrinque plagis tribus nigris, una laterali prope basin, secunda transversa, mediana, tertia transversa inter medium apicemque; apicibus oblique truncatis, disco ($\male$) utrinque paullo pone basim spina parva recurva armato, ($\female$) inerme Long.* $\male$ 8; $\female$ 9.5-11 mm.

Tikekee in Pegu ($\male$); Carin (district of Chebà) ($\female$).

Clothed with a dull cinereous-grey pubescence. The elytra with black markings consisting, on each, of: 1st a short basal vitta which unites, behind the scutellum, with its fellow to form a triangular or y-shaped figure; 2nd an irregular patch placed below, and extending a little behind, the shoulder; 3rd a transverse plaga or fascia situated at about the middle; and 4th a somewhat similar plaga between the middle and the apex. The prothorax is longer than broad, somewhat constricted towards the base, with the sides feebly rounded in the middle, and each furnished with a minute and almost imperceptible tooth; the disk exhibits some minute, asperate, scattered punctures. The elytra appear rather thickly punctured towards the base and as far as the middle, but are almost impunctate towards the apex. The disk of each elytron in the male is armed at some distance from the base with a small backwardly directed spine.

In three examples from the Carin Hills, which I take to be females of this species, on account of their very close agreement in colour and markings with the male type, this discal spine of the elytra is wanting; the disk of the prothorax also does not show the asperate punctures which are present in the male.

The species is undoubtedly congeneric with *R. spinosula*, Pasc.

but may be readily distinguished by certain structural characters as well as by its colour and markings. Its elytra are obliquely truncate and not emarginate nor mucronate at their apices; nor are they asperate at the base. In view of the nature of the sexual differences in the present species, I think it not unlikely that *R. simplex*, Pasc. may prove to be the female of *R. spinosula*, Pasc.

196. **Rondibilis vittata**, sp. n. *Precedenti valde affinis, sed elytrorum sutura lateribusque nigris, disco medio utrinque vitta lata cinerea. Long. 8 mm.*

Carin Mts. (district of Chebà), alt. 900-1100 m. One female example.

Very closely allied to the preceding, differing only in the colour of the elytral pubescence, which is black, with a tolerably broad cinereous band extending along the middle of each elytron from the base to the apex. With a large series of examples it may be shown to be only a variety of *R. plagiata*, but for the present it may be considered distinct.

197. **Ostedes**, sp.

Kawkareet in Tenasserim. One example.

198. **Exocentrus alboseriatus**, sp. n. *Rufo-testaceus, erecte setosus; corpore subtus pedibus antennisque piceo-nigris, his articulis 3° et 4° basi cinereis; elytris maculis parvis numerosis, albo-pubescentibus in seriebus longitudinalibus ordinatis. Long. $5\,^1/_2$-8 mm.*

Thagatà in Tenasserim.

Reddish testaceous. Prothorax transverse, with the lateral spines directed obliquely backwards. Elytra reddish-brown, rather closely punctured on the anterior half, with the punctures becoming sparser posteriorly; each elytron with about five rows of short tufts of white hairs. Antennae black, setose, with the third and fourth joints ashy at the base. Legs pitchy black, with a faint greyish pubescence. Body underneath, the prothorax excepted, pitchy black.

199. **Exocentrus fumosus**, sp. n. *Niger, erecte setosus; elytris pone medium fascia transversa irregulari fulvo-brunnea. Long. $5\,^1/_2$-7 mm.*

Tenasserim, Thagatà and Mt. Mooleyit.

Dull black; the elytra with an irregular transverse testaceous band which is covered with fulvous-brown pubescence and which is placed about halfway between the middle and the apex. Prothorax transverse, minutely and closely punctulate above, black, with the anterior and posterior borders obscurely testaceous; the lateral spines pointing almost directly backwards. Elytra very closely punctured on the anterior half or two-thirds, very sparsely punctured posteriorly. Legs and underside of body brownish-black. The whole body, legs, and antennae furnished with rather long and erect black setae.

200. **Exocentrus**, sp.
Rangoon. One example.

201. **Glenea aeolis**, Thoms. Rev. et Mag. de Zool., 1879, p. 19.
Carin (district of Chebà); alt. 900-1100 m.

202. **Glenea Laodice**, Thoms. Rev. et Mag. de Zool., 1879, p. 15.
Carin (district of Chebà); alt. 900-1100 m.

203. **Glenea anona**, Pasc. var. — *G. anona*, Pasc., Longic. Malayana, p. 393.

The last three or four joints of the antennae are dark-brown, with a cinereous pubescence. In the type of the species these joints are white. One example of this variety was taken at Thagatà in Tenasserim.

204. **Glenea pulchella** ([1]), Thoms. — Essai d'une Class. des Cerambycides, p. 58.

Teinzò, Bhamo, Shwegoo in Upper Burma, Thagatà on the Mt. Mooleyit in Tenasserim and on the Carin Mts. (district of Chebà, alt. 900-1100 m.): also N. India.

205. **Glenea Diana**, Thoms. — Syst. Ceramb. Appendix, p. 561.
Glenea bimaculiceps, Gahan, Trans. Ent. Soc. Lond. 1889, p. 215.

Carin Mts (district of Chebà, alt. 900-1100 m.): also Moulmein and Rangoon (B. M. collection), and Assam (see Thoms. l. c.).

([1]) The specimens from Borneo described under this name by Mr. Pascoe (Longicornia Malayana, p. 370) do not belong to the species, and may be considered to form a variety of *Nicanor* Pasc.

206. **Glenea modica**, Gahan. — Trans. Ent. Soc. Lond. 1889, p. 217.

Two examples taken at Thagatà in Tenasserim.

207. **Glenea spilota**, Thoms. — Essai d'une Class. des Ceramb. p. 58.

Shwegoo and Teinzo in Upper Burma.

208. **Glenea arithmetica**, Thoms. var.

G. arithmetica, Thoms. Archiv. Entom. I (1857), p. 143.

One example taken at Mt. Mooleyit. Alt. 1600 m.

This variety has two greyish-white spots on each elytron in addition to the sutural band. One spot is placed anteriorly, close to, but not in contact with, the sutural band; the second spot, at about the middle of the length of the elytron, touches the sutural band. The external apical angles of the elytra are less distinctly toothed than in Ceylonese examples. Beyond these small differences, the Burmese example before me seems to exhibit no decided characters by which it could be considered specifically distinct from the examples from Ceylon which I have named from comparison with the type specimen.

209. **Glenea posticata**, sp. n. *Atra; capitis fronte (supra medio excepto) albo-flavescente, prothorace albo-flavescente, macula media dorsali et macula laterali utrinque atris; scutello albo-flavescente; elytris dorso griseis, fascia transversa paullo ante apicem albo-flavescente, basi apiceque et lateribus deflexis nigris. Long.* 11 mm.

One example taken at the district of Chebà.

Cheeks and front of the head below with a yellowish white pubescence which extends upwards on each side as far as the insertion of the antennae. Prothorax with a yellowish-white pubescence; with a transversely-oval black spot on the middle of the disk, and a similar spot low down on each side. Elytra black at the sides and at the base and apex, grey above, with a transverse yellowish white band placed a little before the apex. Sides of the abdomen and hind breast and an oblique band on each side of the metasternum, greyish-white.

This species seems to come nearest to *G. indiana,* Thoms. and its allies.

210. **Glenea Gestroi**, sp. n. (Pl. I, fig. 14). *Rufo-brunnea, citrino ornata; capitis fronte (medio excepto), vitta lata laterali prothoracis, lateribus pectoris, basi elytrorum maculisque magnis duabus ad medium, sutura conjunctis, maculis duabus ante apicem, et maculis duabus parvis ad apicem extremum citrinis; antennis articulis duobus basalibus nigris, ceteris rufo-brunneis. Long.* 11 mm.

Bhamò in Upper Burma. One example.

Reddish brown, thickly and strongly punctured. Head with lemon-yellow pubescence in front (along the middle excepted). Prothorax with a lemon-yellow band on each side; the upper borders of these bands are sub-parallel, so that the median dorsal space enclosed between them is nearly oblong in shape. Elytra with a basal transverse band, two large rounded spots at the middle which touch one another at the suture, two smaller spots before the apex, and two very small spots placed at the extreme apex, lemon-yellow. Sides of the breast similarly coloured. Apices of the elytra with the inner angles slightly, the outer angles strongly and distinctly, spined. Legs yellowish brown.

This species somewhat resembles *Glenea vesta*, Pasc. but may may be distinguished by its paler colour; the oblong form of the dorsal median brown space of the prothorax; the presence of two distinct spots, conjoined at the suture, which are placed at the middle of the elytra, and of a small transverse or slightly oblique spot at the extreme apex of each elytron. The abdomen moreover is almost entirely reddish brown underneath.

211. **Glenea cancellata**, Thoms. — Syst. Ceramb. Appendix, p. 565.

Taken at Bhamò and Shwegoo in Upper Burma, and at Thagatà in Tenasserim; occurs also in Siam, and at Sylhet and Darjeeling in N. India.

212. **Glenea nigrolineata**, sp. n. (Pl. I, fig. 15). *Testacea, fulvo-brunneo-pubescens; prothoracis dorso vittis duabus grisescentibus nigro-limbatis; lateribus utrisque obsolete bivittatis; elytris fulvo-brunneis, utrisque humero, maculis duabus conjunctis ante medium, vittis duabus vel tribus brevibus inter medium apicemque et macula*

transversa ad apicem nigris; antennis nigris; pronoto medio carinato; elytris utrisque lateraliter bicarinatis, et dorso prope basin unicarinatis; apicibus utrisque bispinosis. Long. 14. *lat.* 4 *mm.*
Upper Burma, Bhamó and Catcin Cauri.

Testaceous, clothed with a fulvous-brown pubescence. Prothorax with two fulvous-grey vittae on each side of the median dorsal carina; these vittae are bounded by black lines four in number, and extend on to the occiput of the head; the sides of the prothorax have each two indistinct greyish fulvous bands. Elytra fulvous-brown, with black markings which include a narrow glabrous spot on each shoulder, two conjoined spots on each elytron just before the middle, three short vittae between the middle and apex (the two outer of which are longer than the innermost and unite posteriorly) and a transverse spot or band at the apex. Each elytron has two carinae at the side, and one on the disk; the latter carina is distinct enough near the base but becomes obsolete before the middle. Elytra rather sparsely but distinctly punctured; prothorax with a few indistint punctures. Legs and body underneath testaceous, with a fulvous brown pubescence; the tarsi and the inner side of the hind femora somewhat blackish; claws of the male appendiculate; first joint of the four anterior tarsi in the same sex rather strongly dilated. Antennae black, with a slight greyish pubescence underneath.

213. **Glenea nigromaculata**, Thoms. — Systema Cerambycidarum. Appendix, p. 566.

Glenea? Amelia Gahan, Trans. Ent. Soc. Lond. 1889, p. 224.

Two female examples, one taken at Teinzo in Upper Burma, the other at the district of Chebà, Carin Mts.

The females possess also that peculiar character of the claws of the tarsi which I have already pointed out (loc. supra cit.) as existing in the males. The females are to be distinguished from the males by their somewhat larger size, shorter antennae, and by having the last abdominal segment broader, and impressed along the middle of its ventral surface by a feeble groove.

214. Glenea indiana, Thoms. Synops. p. 141.

Palon; Carin Mts. (district of Chebà), 900-1000 m. alt.

215. Glenea cardinalis, Thoms. Class. Long. p. 344.
Thagata, IV, 87.

216. Stibara rufina, Pasc. — *Glenea rufina*, Pasc. Trans. Ent. Soc. Lond. Ser. 2, Vol. IV, p. 259. — *Stibara obsoleta*, Thoms. Essai d'une Class. des Cerambycides, p. 60.

There is a considerable amount of variation, in the coloration of the prothorax amongst the examples which I refer to this species. In some the prothorax is entirely fulvous, in some there is a small greyish-black spot on each side of the disk; in others these spots elongate to become vittae, while in a fourth set of examples the upper side of the prothorax is almost entirely greyish-black. The elytra in all are greyish-black, with a dorsal fulvous vitta, narrowing from the base, on each elytron. This vitta in some examples is exceedingly short, forming in fact a mere basal spot; while in others it extends almost to the apex of the elytra. The punctuation of the elytra in one series of examples appears to be somewhat stronger than in another set; but as this character also is variable, and the difference slight, I am forced to regard all the examples as belonging to a single variable species.

Examples were taken by M. Fea at Bhamò in Upper Burma, at Palon in Pegu and at the district of Chebà, Carin Mts. — The British Museum collection contains examples from Assam Siam, Burma, and Perak.

217. Stibara tetraspilota, Hope. — Trans. Linn. Soc. XVIII (1841), p. 598, Pl. 40, fig. 8; Ann. & Mag. Nat. Hist. VI (1841), p. 300.

Carin Mts. (district of Chebà). One example.

The species appears to be not uncommon in Northern India (Darjeeling, Assam etc.).

218. Nupserha nigriceps, sp. n. *Fulvo-testacea; capite, antennis, elytrorum abdominisque apicibus, tibiis posticis et tarsis nigris; prothorace quam longiori paullo latiori; antennis quam corpore longioribus. Long.* 13-14 *mm.*

Carin Mts. (Chebà district).

Fulvous or yellowish-testaceous. Head, antennae, the apical fifth or sixth of the elytra, the apical half of the last abdominal segment, the posterior tibiae and tarsi black. Prothorax with its sides subparallel, or very slightly rounded in the middle between the anterior and posterior transverse grooves; the width of the prothorax slightly exceeds its length, the difference being more marked in the female than in the male. Elytra rather strongly and thickly punctured as far as the apical black patch, with the punctures arranged more or less regularly in close longitudinal rows. The disk of the elytra is slightly flattened and depressed a little below the base, and has an obsoletely raised line towards its outer edge on each side; the sides of the elytra have each a rather indistinct longitudinal carina; the apices are truncate in a slightly oblique direction, the inner angles being less prolonged than the outer. The antennae in both sexes are longer than the body, those of the male exceeding it by about the last three joints. The last ventral abdominal segment of the male has a rather deep semi-oval depression along its hinder half.

From *N. ustulata,* Erichs. which it very closely resembles, the present species may be distinguished by its black-head, slightly longer antennae, and the less distinct lateral carinae of its elytra.

219. **Nupserha ventralis,** sp. n. *Capite rufo-fulvo, pronoto pallide fulvo, elytris flavo-testaceis; corpore subtus pedibusque (femoribus anticis nonnihil testaceis, exceptis) et antennis nigris; his quam corpore paullo longioribus; elytris lateraliter utrinque acute unicarinatis; apicibus sub-oblique truncatis, angulis externis magis productis et spinosis. Long.* 10-12 *mm.*

Carin Mts. (Chebà district), alt. 900-1100 m.

Prothorax rather broader than long, feebly swollen on each side just in front of the basal transverse groove; the disk with a faint median raised line. The pale fulvous upper side of the prothorax, clearly marked off by a straight border, along the middle of each side, from the black underside of the prothorax.

Elytra with a sharp carina on each side, which extends up to the external apical angle; at this angle the apices are spinose, at the sutural angle dentate.

The species seems to be nearly related to *Nupserha pallidipennis*, Redtenb. *(Phytoecia)*.

220. **Nupserha variabilis**, sp. n. *Capite prothoraceque rufo-fulvis, sat dense punctatis; prothorace quam longiori fere sesqui-latiori, lateribus in medio fortiter rotundatis vel sub-tuberosis; elytris, postice exceptis, sat dense sublineatimque punctatis, lateraliter unicarinatis, dorso utrinque uni-costato; dimidio antico elytrorum nigrescente, griseo sat dense pubescente, dimidio postico (fascia transversa apicali nigra excepta) testaceo, sub-glabro; apicibus truncatis, angulis leviter dentatis; corpore subtus pedibusque fulvis, his tibiis posticis apice et tarsis omnibus nigrescentibus; antennis nigris, articulo* 3° *quam* 4° *vix longiori. Long.* 12-15. *Lat.* $3\text{-}4\,^1/_2$ *mm.*

Variat. *Dimidio basali elytrorum plus minusve testaceo.*

Siam and Tenasserim (British Museum collection).

Taken by Signor Fea at Bhamò in Upper Burma, at Carin Mts. 400-600 m.; at Palon, at Rangoon and at Thagatà.

This species is subject to a certain amount of variation in the extent to which the black coloration spreads over the basal half of the elytra. In a male example taken at Bhamò the black on the basal half of the elytra is confined to a small longitudinal patch on, and behind, each shoulder. In a female specimen from Thagatà the anterior sixth of the elytra is testaceous, the grey-black colour forming a zone between this and the middle. In those examples in which the grey-black is most extensive, there is still a small testaceous patch left at the extreme basal margin. In other respects the species appears to be constant enough. The transverse black fascia at the apex of the elytra is always very distinct. The lateral carina passing back from each shoulder is sharp and prominent, but does not extend quite up to the apex; the dorsal costa on each side limits externally a rather broad shallow depression which lies between it and the suture. The black antennae are about equal in length to the body in the female, a little longer in the male. Tarsal claws of both sexes appendiculate. Last ventral segment of the male with a somewhat oval depression along the middle.

221. **Nupserha antennata**, sp. n. *Fulvescens, griseo-subtilissime pubescens; elytris flavo-testaceis, fascia apicali nigra; prothorace vix punctato, quam longiori multo latiori, antice posticeque transversim sulcato, lateraliter in medio sub-tuberoso; elytris dense fortiterque punctatis, lateraliter uni-carinatis; antennis quam corpore longioribus, nigris, articulis intermediis basi plus minusve fusco-testaceis, articulo* 4° *quam* 3° *longiori. Long.* 12-15. *Lat.* 3-4 $1/2$ *mm.*

Carin Mts. (Chebá district), and Upper Burma (Bhamò).

Fulvous, with a faint greyish pubescence. Elytra with the apical fourth or fifth black, the rest yellowish testaceous: closely and rather strongly punctured up to a little beyond the middle; with a well defined carina on each side which does not quite reach to the apex; the disk not depressed and scarcely even flattened along the middle, without a distinct costa; apices truncate in a slightly oblique direction, with the angles very feebly mucronate. Body underneath fulvous, with a small spot on the side of the prothorax, the sides of the meso-and metathorax, and tip of abdomen blackish. Legs fulvous, with all the tarsi, the hind tibiae, and, to a slight extent, the middle and hind femora blackish-brown.

In its general build and in the form of its prothorax this species has a close resemblance to *N. variabilis*, and to an undescribed species from North India; its antennae are, however, quite different, being slenderer and longer, and having the fourth joint equal in length to the second and third united.

222. **Nupserha dubia**, n. sp. *Fulvescens; capite prothoraceque sat dense punctatis; hoc quam longiori vix latiori, lateribus medio leviter rotundatis; elytris sat fortiter crebreque punctatis, punctis lineatim ordinatis, postice evanescentibus, lateribus anguste et apicibus nigro-fuscis, apicibus oblique truncatis, angulis sub-acutis vix dentatis; corpore subtus fulvo, mesothoracis episternis metatho-racisque lateribus nigro-plagiatis; tarsis omnibus et tibiis posticis nigro-fuscis; antennis quam corpore longioribus, articulis* 1° *ad*

4um *nigris, ceteris basi obscure testaceis, apice nigro-fuscis.* Long. 10 ¹/₂. Lat. 3 mm.

Carin Mts. (Chebà district), alt. 900-1100 m. Two male examples.

The lateral carina of the elytra is less distinct in this species than in most of the known species of *Nupserha*. In size and coloration it somewhat resembles *N. ustulata*, Erichs., from which it is easily to be distinguished by the character just mentioned as well as by the absence of a distinct dorsal costa from the elytra, the obscurely annulated antennae and other less important characters. The last ventral segment of the male of the present species has a rather deep triangular depression on the posterior third.

Nupserha annulata, Thoms. (*Stibara*) approaches the present species very closely in structural characters, and is to be distinguished from it by its concolorous (yellowish-testaceous) apices of the elytra, the more distinct fuscous colour at the margins of the elytra, which, however, does not reach the base or apex, the fuscous spot or band at each side of the prothorax, and by the almost entire greyish-black colour of the underside of the body.

223. **Nupserha fricator**, Dalm. — *Saperda fricator*, Dalm. Schönherr. Synonymia Insectorum. Appendix, p. 183.

Carin Mts. (Chebà district).

224. **Nupserha quadrioculata**, Thunberg. — Mus. Nat. Acad. Upsalae, IV (1787), p. 57. — *Stibara carinata*, Thoms. Archiv. Entom. I, p. 146.

Bhamò, Upper Burma; and Carin Mts. (Chebà district).

Also occurs in Java, Siam, Tenasserim and North India (Sylhet).

225. **Oberea posticata**, sp. n. *Fulvescens, elytris brunneo-testaceis apice nigro-fuscis; antennis nigris, abdomine apice anguste nigro; prothorace quam latiori paullo longiori, lateribus sub-parallelis; elytris elongatis, crebre punctatis, apicibus sub-oblique truncatis, angulis mucronatis, abdominis segmento ultimo ventrali fortiter excavato; antennis quam corpore paullo longioribus, articulo 3° quam 4° evidenter longiori.* Long. 13-18 mm.

Teinzo in Upper Burma, Meetan and Carin Mts. (district of Chebà). The British Museum has examples from Darjeeling, Nepal, and Sylhet in North India.

This species is to be distinguished from those having a similar coloration, by the deep and rather broad excavation which occupies nearly the whole length of the ventral side of the last abdominal segment in the male. The abdomen, like the apices of the elytra, is usually tipped with brownish black. The hind femora do not reach beyond the middle of the second abdominal segment. From *O. fuscipennis*, Chev., which it most nearly resembles, the species may be distinguished by its somewhat narrower and more elongated prothorax, and the much deeper excavation of the last abdominal segment of the male.

226. **Oberea armata**, sp. n. *Fulvescens; elytris brunneo-testaceis, griseo leviter pubescentibus, versus basin pallidioribus; antennis nigris, tibiis posticis infuscatis; prothorace quam longiori manifeste latiori, dorso medio sub-gibboso; elytris dense punctatis, medio angustatis; segmento primo abdominis pallido, ad apicem medio processu angusto armato (σ), segmento ultimo prope apicem subtus leviter cordato-impresso, apice truncato; femoribus posticis quam segmentis duobus basalibus abdominis paullo longioribus. Long.* 16 mm.

Thagatà, Tenasserim and Carin Mts. (district of Chebà).

Fulvous; elytra somewhat pale brown with a greyish pubescence. Prothorax broader than long, slightly gibbous on the middle of the disk. First segment of the abdomen in the male produced behind into a narrow median process which extends back almost three fourths the length of the following segment. This process is slightly expanded towards its extremity, where also it bears a short and not very dense fringe of fulvous hairs. The last ventral segment of the male is truncate at the apex, just in front of which it bears a feeble heart-shaped depression. The antennae are black, and somewhat shorter than the body; the third joint is scarcely perceptibly longer than the fourth. The hind femora reach a little beyond the hind border of the second abdominal segment.

This species has a close general resemblance, and is probably nearly allied, to *O. fuscipennis*, Chevr. In the latter species the elytra are less pubescent and somewhat more strongly punctured, the last ventral segment of the male has a larger and deeper depression, and the first segment is concolorous and unarmed. The curious abdominal process with which *O. armata* is provided seems to be confined to this species. I do not recollect the occurrence of anything similar to it elsewhere among the Longicorns. A somewhat similar process does, however, occur among certain species of *Galerucinae* (Gen. *Hoplasoma*, Jac.).

227. **Oberea Birmanica**, sp. n. *Fulvescens; elytris lateraliter posticeque sub-infuscatis; pedibus posticis abdomineque fuscis, hoc segmento primo toto et secundo in medio argenteo-sericeis, ceteris medio subtiliter griseo-sericeis; antennis (articulis 1°, 2°que fulvis, exceptis) nigrescentibus, quam corpore brevioribus, articulo 3° quam 4° longiori; prothorace quam longiori distincte latiori, lateribus rotundatis. Long. 16-19 mm.*

Upper Burma, Teinzó and Bhamó.

This species agrees very closely in structural details with *O. curialis*, Pasc., which occurs in Sumatra, Java and Penang. The difference is chiefly one of colour. In *curialis* the elytra are distinctly black, while the head and prothorax are reddish. In the present species the elytra are fulvous, testaceous somewhat infuscate externally and towards the apex; the head and prothorax are yellowish tawny. So far as I can judge from the few specimens of each form that I have seen, these differences remain constant; though possibly the species I have just described may prove to be only a local or pale variety of *O. curialis*.

228. **Oberea sericea**, sp. n. *Capite, prothorace et corpore subtus fuscis, pube griseo-sericea dense obtectis; elytris testaceis, lateraliter posticeque et ad suturam infuscatis, pube griseo-sericea minus dense obtectis; pedibus testaceis, posticis tibiis femorumque apicibus fuscis, antennis nigrescentibus. Long. 13. Lat. 3 mm.*

Carin Mts. (Chebà district). Alt. 400-900 m. One example.

Head, prothorax, scutellum and underside of body with a dense pearl-grey pubescence which has a lustre like that of shot silk, appearing lighter or darker in shade according to the light in which it is viewed. The elytra have a similar but less dense pubescence through which can be seen the colour of the derm, testaceous at the base and for some distance along the middle of each, dark brown posteriorly and along the sides and suture. The prothorax is scarcely broader than it is long, and has its sides almost perfectly straight and parallel. The hind tibiae, the distal half of the hind femora, and all the tarsi are more or less infuscate; the remaining portions of the legs are yellowish testaceous. The hind femora reach beyond the middle of the third abdominal segment.

229. **Oberea modica**, sp. n. *Modice elongata, fulvescens; prothorace quam longiori vix latiori, antice posticeque sulcato, lateribus medio vix rotundatis; elytris supra deplanatis, lineatim punctatis, apicibus lateribusque (his versus basin exceptis) infuscatis; antennis quam corpore longioribus, nigris, vel articulis intermediis basi fusco-testaceis, articulo 4° quam 3° longiori.* Long. 12-14. Lat. 3-3 $^1/_2$ mm.

Bhamò in Upper Burma; also Northern India (B. M. collection).

Fulvous; sides of the elytra (towards the base excepted) and apices dark brown. Antennae black; tarsi and hind tibiae infuscate. Prothorax scarcely broader than long, transversely sulcate anteriorly and posteriorly; with its sides very feebly rounded in the middle and almost parallel. Elytra flattened or slightly depressed along the middle above; closely punctured, with the punctures arranged in more or less regular lines; apices truncate, with the angles scarcely acute. Antennae distinctly longer than the body, black, or with the intermediate joints somewhat brownish testaceous at their bases; the fourth joint almost equalling in length the second and third united. Hind femora reaching to the apex of the third abdominal segment. This species belongs to a small group whose somewhat shortened form brings them into close resemblance to species of *Nupserha*, from which, however, they are to be distinguished by the absence of a distinct lateral carina to the elytra. *O. mar-*

ginella, Bates, which also belongs to this group, may be distinguished from the present species by its black head, the more clearly limited fuscous vitta at each side of the elytra, and by the structure of the last dorsal segment of the female which ends in two sharp teeth between which is a deep but rather narrow emargination. In the present species this plate is rounded at the apex.

230. **Oberea pallidicornis**, sp. n. *Precedenti affinis sed differt elytris medio leviter angustioribus, antennis (articulis* $1°$, $2°que$ *nigris exceptis) pallide testaceis, articulis* 3^1 *ad* 6um *apice angustissime nigris. Long.* 12. *Lat.* $2^1/_2$ *mm.*

Carin Mts. (district of Chebà).

Signor Fea has taken two examples, which though closely allied to, appear to be distinct from the preceding species. They have on the whole a somewhat narrower form: the front of the head in the male is decidedly narrower than in *modica,* and in this sex also the last ventral segment is very feebly or scarcely at all impressed. (In a North Indian example which I take to be the male of *O. modica* this segment has a distinct, though not strongly marked, longitudinal impression).

231. **Oberea brevis**, sp. n. *Fulva, elytris lateraliter posticeque infuscatis, apicibus oblique truncatis fere rotundatis; capite prothoraceque dense punctatis, hoc lateribus fere parallelis; elytris crebre punctatis, punctis postice evanescentibus; antennis articulis* $1°$ *ad* 3um *nigro-fuscis, ceteris obscure testaceis, apicibus fuscis; articulo* $3°$ *quam* $4°$ *vix longiori. Long.* $9^1/_2$. *Lat.* $2^1/_2$ *mm.*

Bhamò in Upper Burma. One female example. Male, without indication of locality, in Brit. Mus. collection.

Fulvous. Sides and hinder part of elytra turning to dark brown. Antennae longer than the body in both sexes with the first three joints black-brown, the remainder more or less testaceous; the third joint barely longer than the fourth. Hind legs not much longer than the middle or anterior pairs; their femora reaching almost to the hind border of the third abdominal segment. Last ventral in the female rather long, and marked with a distinct longitudinal line; the same segment in the male somewhat shorter, emarginate at the apex, and slightly impressed in the middle just before the apex. This is one of the shortest species of *Oberea* known me. In its size and general aspect it resembles some of the European species of *Phytoecia*, and in fact cannot be separated from them by any decided structural characters. On the other hand I see no good reason for separating it generically from such species as *Oberea marginella, modica, sericea* and others.

232. Phytoecia amoena, sp. n. (Pl. I, fig. 16). *Pube sulphurea, nigro ornata, obtecta; antennis griseo-nigris, pedibus testaceis, griseo subtiliter pubescentibus. Long. 9 mm.*

Carin Mts. (Chebá district). Alt. 900-1100 m. One female example.

Clothed with a sulphur-yellow pubescence, with black markings. Head with a black spot on the lower part of the front, and with a triangular spot on the occiput, with a median line uniting the opposed apices of the two spots. Prothorax with its median length about equal to its anterior width, its sides nearly parallel and slightly constricted just before the base; the disk with two elongated black spots placed one on either side of the middle line, the sides each with a small black spot. Elytra each with a submarginal black line passing back from the shoulder along the outer edge of the disk as far as the middle, a line passing from a little behind the base, close to the suture, as far as the middle where it unites with a black spot, and, between this point and the apex, two transverse black spots or fasciae, the posterior of which is much narrower than the anterior, and which are united at their sutural extremities by a narrow black line. Sides of the elytra almost parallel, apices transversely truncate. Body underneath with a sulphur yellow pubescence, which is somewhat less dense along the middle. Legs testaceous, with a faint greyish pubescence.

This species has very much the general aspect of a small *Saperda*. Its position in *Phytoecia* is necessitated by the character of its tarsal claws, which are distinctly appendiculate.

233. **Astathes gibbicollis.** Thoms. Systema Ceramb. p. 559.
Bhamò in Upper Burma.

234. **Astathes dimidiata**, Gory. — *Tetraopes dimidiata*, Gory. Guér. Icon. règne animal., p. 244, t. 45, fig. 3.
Meetan in Tenasserim.

235. **Astathes violaceipennis**, Thoms., Archiv. Ent. I, p. 53.
— *Tetraophthalmus violaceipennis*, Thoms., Archiv. Ent. I, p. 53.
Carin Mts. (district of Chebà).

236. **Chreonoma frontalis**, sp. n. *Testaceo-flava, setosa, postpectore abdomineque et elytrorum apicibus nigro-cyaneis, antennis versus apicem nonnihil infuscatis; capite sparse-, prothoraceque sat dense punctatis; elytris dense punctatis, punctis versus apicem evanescentibus. Long.* 11-13 mm.

Bhamò in Upper Burma. One example. Also North India (Brit. Mus. Coll.).

Yellowish testaceous or fulvous; the elytra somewhat paler; hind-breast, abdomen and the posterior fifth or sixth of the elytra dark metallic blue. Head feebly and sparingly punctured, the front projecting in the middle in the form of a carina or laterally-compressed tubercle, which is less well marked in the female. Prothorax somewhat more thickly and more strongly punctured; the disk with a tolerably distinct, though not sharply limited, central umbone or swelling which is somewhat rounded or transversely-oval in shape. Elytra rather thickly punctured over the whole of the yellow area; less thickly and not so manifestly punctured on the dark-blue apical portion. In addition to these larger setigerous punctures, the interstices between them exhibit a very fine punctuation.

This species somewhat resembles *Chreonoma dilecta*, Newm. *(Phœa)* but the latter has the antennae almost entirely black, the whole of the underside of the body fulvous testaceous, the apex of the elytra piceous; the prothorax also is less thickly punctured, is very feeble raised on the middle of the disk, and the lateral tubercles are almost obsolete; the head of the female has a raised line along the middle of the lower part of the front. *Chreonoma melanura*, Pasc., seems, from the description, to have a similar style of coloration to the present species; but may be distinguished by its dark hind legs, and by its different punctuation.

ADDENDA.

The following species had been held over in the hope that a comparison with types in Pascoe's collection might lead to their identification. They seem, however, to be new; but only those represented by well preserved specimens are described.

237. **Gen.? sp.**
Thagatà in Tenasserim.

This, apparently new genus, is represented by only a single somewhat imperfect specimen. It has the shortened elytra of *Epania* or *Necydalis*; but this character is accompanied by a roundness and fullness of the prothorax which is more suggestive of the *Clytides*. The femora are strongly clavate, but less abruptly so than in *Epania*. The state of the specimen does not permit me to examine sufficiently the structure of the cotyloid cavities.

238. **Demonax quadricolor**, sp. n. *Capite prothorace, corpore subtus, pedibusque et antennis cinereis, episternis meso-, metathoracisque albescentibus; elytris fasciis pubescentibus ornatis, prima basali cinerea, maculam nigram oblique elongatam includente, secunda angusta, nigra, valde obliqua, tertia griseo-fulva triangulari, vix pone medium posita, quarta fulvo-brunnea transversa, quinta griseo-fulva, lata, transversa, apicali; antennis (♀) medium elytrorum haud superantibus, articulis $3°$, $4°que$ extus breviter spinosis.* Long. $10\,^{1}/_{2}$ mm.

Hab. Mount Mooleyit in Tenasserim. Alt. 1000-1900 m.

This species somewhat closely resembles *D. salutarius*, Pasc. and other species in the pattern of its markings, but it may be distinguished by the colour of the three posterior bands of the elytra; the middle one of these is fulvous-brown in colour, while the submedian band, immediately in front of it, and the apical band, just behind it, are both of a somewhat greyish-tawny shade. The third and fourth joints of the antennae are very briefly spined at the outer distal angle.

239. **Blepephaeus parvicollis**, sp. n. *Pube fulvo-brunnea sericea sat dense vestitus; elytris albido-cinerascentibus, plaga triangulari, ad basin et plaga subtriangulari et sub-elongata utrinque brunneo-velutinis; copite prothoraceque fere impunctatis, hoc lateraliter utrinque sat valde spinoso, supra tuberculis tribus parvis, et postice granulis nigris paucis obtecto; elytris versus basin sparse asperato-punctatis; apicibus rotundatis; mesosterno medio sat distincte tuberculato, tibiis intermediis prope medium leviter oblique emarginatis.* Long. 23. Lat. 8 mm.

Hab. Catcin Cauri in Burma.

The relatively great width of the elytra as compared with the head or prothorax (the width of the elytra across the base being about twice that of the base of the prothorax) gives to this species a characteristic appearance. Some of its remaining characters do not quite agree with those of the genus *Blepephaeus*. The mesosternum has a small but distinct tubercle at about the middle of its length; the middle tibiae bear an evident, though feeble, oblique groove just below the middle.

240. **Batocera**, sp.

Palon in Pegu. One somewhat rubbed specimen.

241. **Pterolophia chebana**, sp. n. ♀. *Fusca, pube fulvo-grisea, fusco-mixta sub-maculatim vestita; antennis medium elytrorum paullo superantibus, articulis 3° et 5° ad 11um basi anguste fulvo-griseis, 4° fulvo-griseo, apice fusco; capite prothoraceque sparse punctatis; elytris dense fortiterque punctatis, punctis sub-oblongis, dorso sub-planis haud cristatis.* Long.

Hab. Carin (district of Chebà; alt. 900-1100 m.). One female example.

Head very sparsely punctured, clothed with a not very dense greyish tawny pubescence. Prothorax a little more strongly and more thickly punctured, with a similar greyish tawny pubescence, which is more widely interrupted by dark brown. It is only slightly convex above. Elytra with a fulvous grey pubescence, a good deal interrupted with dark brown towards the base, and crossed by a broad but rather indistinct dark brown band just behind the middle. They are slightly flattened above, and are thickly punctured with rather large and sub-oblong punctures, which are more distinctly seen on the less pubescent dark brown areas; they are strongly enough and regularly declivous behind, with the apices rounded. The antennae of the female are about equal in length to three-fourths of the body; the fourth joint is widely, the other joints (first, faintly fulvous excepted) narrowly covered with fulvous-grey pubescence at the base.

This species is more flattened and more even above, and the vertical height of the elytra consequently less than is the case with the great majority of the species placed in the genus *Pterolophia* (= *Praonetha*).

242. **Pterolophia bimaculata**, sp. n. *Sub-angusta; capite fulvo-pubescente, supra fusco-bipunctato; prothorace griseo, supra bituberculato; elytris sat dense punctatis, pube grisea, fulvescente paullo mixta, sat dense vestitis; utrisque basi breviter haud distincte cristatis, postice sat gradatim declivis et macula elongata paullo arcuata fusca instructis, apicibus sub-truncatis; pedibus supra fulvo-brunneis, subtus cinereis; antennis (♀) medium elytrorum paullo superantibus. Long. 9. Lat. 3 mm.*

Head with a fulvous-brown pubescence; vertex with two small sub-glabrous black spots. Prothorax grey; the disk with two obtuse tubercles. Elytra grey mixed with fulvous; each, between the middle and apex, bears an elongate dark brown spot, the anterior end of which is somewhat broader and turned in obliquely towards the suture. The legs are clothed with a fulvous-brown pubescence on the upper, and with a cinereous pubescence on the underside.

This species may be recognized by its rather narrow form, its dorsally tubercled prothorax, its posteriorly gradually declivous elytra; and especially by the two post-median dark brown elytral spots, of which that on the right side has somewhat the shape of a comma.

243. **Hyllisia**, sp.

Carin Chebà. One example.

244. **Oberea**, sp.

Bhamò in Upper Burma. One example.

EXPLANATION OF PLATE

Fig. 1. Tetraommatus insignis.
» 2. Dymasius fulvescens.
» 3. Ibidionidum Corbetti.
» 4. Xylotrechus Gestroi.
» 5. Caloclytus ludens.
» 6. Demonax literatus.
» 7. » reticollis.
» 8. Polyphida Feae.
» 9. Agelasta nigromaculata.
» 10. Mesolophus humeralis.
» 11. Niphona princeps.
» 12. Pterolophia alboplagiata.
» 13. Estigmenida variabilis.
» 14. Glenea Gestroi.
» 15. » nigrolineata.
» 16. Phytoecia amœna.

Estratto dagli Annali del Museo Civico di Storia Naturale di Genova
Serie 2.ª, Vol. XIV (XXXIV) 9-10 Aprile 1894.

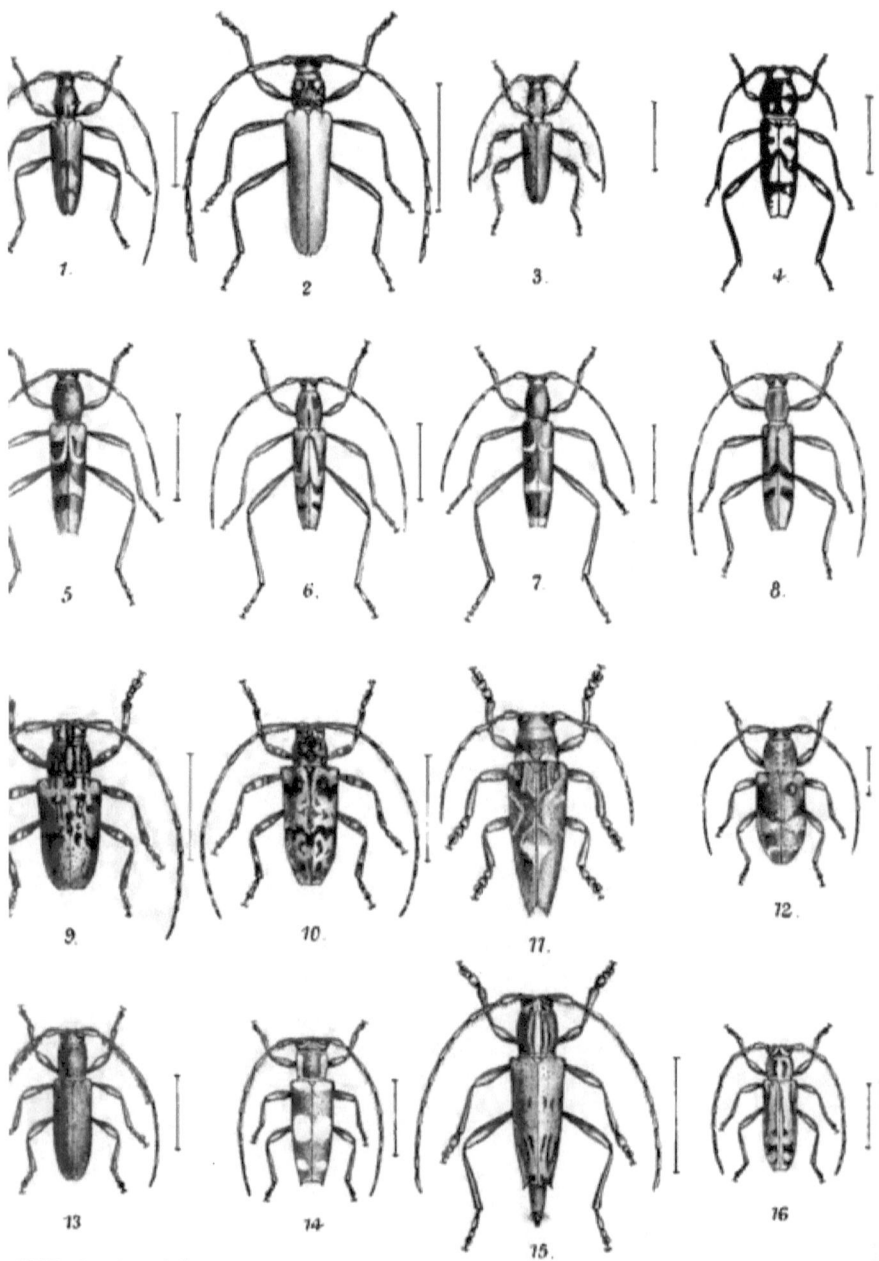

Annali del Museo Civico Ser 2ᵈᵃ Vol XIV 1894.   Tav I

New Longicornia from Burma

www.ingramcontent.com/pod-product-compliance
Lightning Source LLC
Chambersburg PA
CBHW020153170426
43199CB00010B/1020